Energy Resources in Colorado

Other Titles of Interest

The Geopolitics of Energy, Melvin A. Conant and Fern Racine Gold

U.S.-Canadian Energy Trade: A Study of Changing Relationships, Helmut Frank and John T. Schanz

Public Utility Rate Making in an Energy-Conscious Environment, edited by Werner Sichel

Coal Surface Mining: The Impacts of Reclamation, edited by James E. Rowe

Where We Agree: Report of the National Coal Policy Project, edited by Francis X. Murray

Applied Mineral Exploration With Special Reference to Uranium, Robert V. Bailey and Milton O. Childers

Energy Resources in Colorado: Coal, Oil Shale, and Uranium
Jerome G. Morse

As the rate of energy use far outstrips the estimate of what resources may be available, other fuels are needed to supplement dwindling supplies of oil and gas. Three such alternatives are coal, oil shale, and uranium, all of which exist in potential abundance in Colorado. However, large-scale production of these resources will result in complex interactions with the state's social and physical environment, and the technologies and issues involved must be fully identified and understood before full-scale development takes place.

This book provides current information on coal, oil shale, and uranium resources in the state, as well as on the potential impact of the technology associated with their development. Initially prepared as three separate volumes intended to inform policymakers in the legislative and executive branches of state government, this single volume has been expanded and brought up to date for a broader audience.

Jerome G. Morse, formerly deputy director for technology at the Colorado Energy Research Institute (CERI), is now adjunct associate professor of physics at the Colorado School of Mines and consultant to CERI.

This book was prepared under the auspices of
The Colorado Energy Research Institute, Golden, Colorado

Energy Resources in Colorado: Coal, Oil Shale, and Uranium

Jerome G. Morse
research assistance provided by
R. Michael Stanwood

Routledge
Taylor & Francis Group
LONDON AND NEW YORK

First published 1979 by Westview Press

Published 2018 by Routledge
52 Vanderbilt Avenue, New York, NY 10017
2 Park Square, Milton Park, Abingdon, Oxon OX14 4RN

Routledge is an imprint of the Taylor & Francis Group, an informa business

Library of Congress Catalog Card Number: 79-4850

ISBN 13: 978-0-367-02109-2 (hbk)

To Lauren, Dave and Afton

Contents

Page

Page

Tables

Page

Page

Figures

Plates

Foreword

Development of the extensive energy resources of Colorado--particularly coal, oil shale and uranium--is rapidly changing the face of the State. The rate and direction of that change is of great concern to all of us, particularly those who must formulate and make policy decisions, whether in the public or private sectors.

The Colorado Energy Research Institute is concerned with improving the quality of energy policy analysis and decision-making in the State by providing a broad range of scientific, technological and economic research information about energy development and its impacts. The Institute is also concerned with increasing the level of public understanding about energy resources and issues.

Energy Resources in Colorado is significant because it is responsive to both these goals. It is designed to address the information needs of policy-oriented generalists who are concerned with the major energy development issues facing Colorado, as well as energy specialists who may not be aware of all facets of resource development. It represents an effort on the part of the Institute--the first in a series--to provide a generally nontechnical yet comprehensive picture of those energy resources which are undergoing rapid development in the State. Hopefully, it will become a well-used tool for understanding the major issues that are involved in energy development.

Energy is a force that drives us all. As a highly industrialized nation we must be concerned about the what, where, why, how, and when of energy development, and the types of impacts that come about from that development. And, of course, how energy development will change Colorado is of concern to us all.

Martin D. Robbins
Director
Colorado Energy
Research Institute

Preface

The State of Colorado is endowed with vast scenic beauty, long recognized as a fuel to enhance the spirit of resident and visitor alike. In a similar sense, Colorado is endowed with large deposits of low-sulfur coal (fourth in U.S.), oil shale (first in U.S.) and uranium (fifth in U.S.), fuel sources for our nation's energy requirements. These energy resources, coal, oil shale, and uranium, exist in sufficiently extensive amounts to be subjects of strong developmental interest from outside the State's borders. We anticipate that major development of these resources will happen, placing great pressure on the land, the water and the people of Colorado. However, with planning and guidance by our leadership in the executive and legislative branches of State government, energy development can occur with minimum environmental degradation and socioeconomic impact.

Just as one prepares for some great cataclysmic force about to erupt, it is possible--in an engineering sense--to control and channel-off this potentially destructive force into productive pathways and so reduce its effects. The "finger-in-the-dike" response for controlling the onrushing tide of degradation, except in legend, has never been viable, and will not suffice for Colorado. A thoughtful and studied consideration of the facts and issues related to energy development is prerequisite to judging alternatives for action.

This present volume is a revision, updating, and consolidation of three prior reports titled the <u>Colorado Energy Resources Handbooks</u>, <u>Volume 1</u> - <u>Coal</u>, <u>Volume 2</u> - <u>Oil Shale</u>, and <u>Volume 3</u> - <u>Uranium</u>. Then, as now, the overall intent of writing about these energy resources is to provide policy-makers in Colorado, as well as elsewhere, with current and factual data about resource development and its related effects. The nature of technology associated with energy resource development together with an awareness of the options for mitigation of the problems of development is intended to provide State leaders with information for effective decisions.

The scope of the book is broad. It attempts to provide a comprehensive review of the processes that go into the development and utilization of coal, oil shale and uranium for the production of energy. In some cases, this review of processes has been necessarily general, but the purpose has always been to assemble enough information and data to serve as a foundation for more detailed study if desired. Data cited are drawn largely from available published sources.

J. G. Morse
Golden, Colorado

Acknowledgments

The author is sincerely indebted to Joseph V. Guerrero, Colorado Energy, for his continued guidance, encouragement and excellent editorial judgment in preparation of this book. In addition, much thanks goes to:

1) Art Lees, P.E.; John S. Gilmore, DRI; John S. Hutchins, Energy Development Consultants; Joan E. Martin, Colorado Energy for technical review.

2) Glennda Mitch, Debra Titlow, David L. Bucknam for editorial assistance.

3) Pam Davis and Barbara LaSasso for typing many drafts.

4) Ruby Fulk for typing the final draft.

5) Jean Moody for preparation of illustrations.

6) M.S. Curtin, D.H. Hebb, R. Cattany and A. Anderson - all formerly of the Colorado School of Mines and co-authors of prior reports.

7) Cameron Engineers, Energy Fuels, Gulf Oil (Pittsburgh-Midway), Colorado Westmoreland, Rio Blanco, Occidental Oil Shale, Paraho, Union Carbide, Cotter Corp., Behrent Engineering for supplying photos or illustrations.

Part I

Coal

1 Nature of the Resource

COAL: CHARACTERISTICS AND CLASSIFICATIONS

Coal, like oil and natural gas, is a fossil fuel. However, unlike oil and natural gas which were formed from the remains of plants and animals, coal was formed from plant residues alone. Ten million to 360 million years ago, when the process first began, mud and water first prevented these plants from completely decaying (which would have released their stored energy). As it covered the plants, the mud and water gradually built up layer after layer of partially decayed plant into thick deposits.

As time passed, the deposits of partially decayed plants formed a spongy substance called peat which in turn became further buried by large deposits of sediments. During the process of decaying, the plants released large quantities of volatile matter, leaving behind carbonaceous deposits which, under pressure and high temperature formed coal, or lignite (Federal Energy Administration, 1976a, p. 3).

Ranging in color from dark brown to black, coal contains more than 50 percent by weight and more than 70 percent by volume of carbonaceous materials. Besides carbon, coal contains small amounts of oxygen, nitrogen, and other constituent elements. The proportions of these elements, as well as other physical characteristics of coal, will vary widely; they form the basis for coal classification by rank (percentage of fixed carbon and heat content, calculated on a dry mineral-free basis) and by grade or quality (content of sulfur, ash, and other deleterious constituents). Typical ranges of coal values are listed in Table 1.1. Coal with high moisture, sulfur, and ash contents is undesirable for burning as well as for environmental reasons. A high percentage of volatile matter and fixed carbon content provide most of the energy in the combustion of coal.

Although many useful terms are explained in the Glossary (see p. 373), it may be helpful to provide several definitions

Table 1.1

RANK OF COALS

Rank	Class	Fixed Carbon	Volatiles	Moisture	Heat Values
		(%)	(%)	(%)	(Btu/lb)
I	Anthracite	85-90	5-15	5-15	14,000
II	Bituminous	45-85	15-35	5-15	12-15,000
III	Subbituminous	35-45	30-40	20-30	8-11,000
IV	Lignite	25-30	15-30	40-50	6-7,500

Ash content ranges from 2.5 - 32.6% (8.9% average for all coals)

Sulfur content ranges from 0.3 - 7.7% (1.9% average for all coals)

Source: Averitt, 1975, p. 17.

here: Coal resources represent the total amount of coal in the ground within specific limits of coal seam overburden thickness. This includes both identified and undiscovered coal deposits. Coal reserves, on the other hand, are limited to identified resources, and are divided into three categories: measured (based on closely spaced observation), indicated (based partly on specific observation and partly on reasonable geologic projection), and inferred (based on an assumed continuity of coal beds containing measured and indicated reserves).

All reserves are considered economically recoverable using currently available technology. The term reserve base (also called "demonstrated reserves") refers only to measured and indicated reserves, and represents the best estimate of coal resources known in location, quantity and quality (Averitt, 1975, pp. 105-106). The relationship of resources and reserves is illustrated in Figure 1.1.

One important distinction between grades of coal is the difference between coking (metallurgical grade), and noncoking (steam grade) coals. Coking coal is premium quality coal because, when heated to a minimum of 1500° F in the absence of air, it will produce a hard cellular carbon residue (coke) (U.S. Bureau of Mines, 1976, p. 159). Coke plays a major role in the manufacturing of steel. Steam coal, on the other hand, is normally used for the generation of electricity and is by far the more commonly found of the two grades.

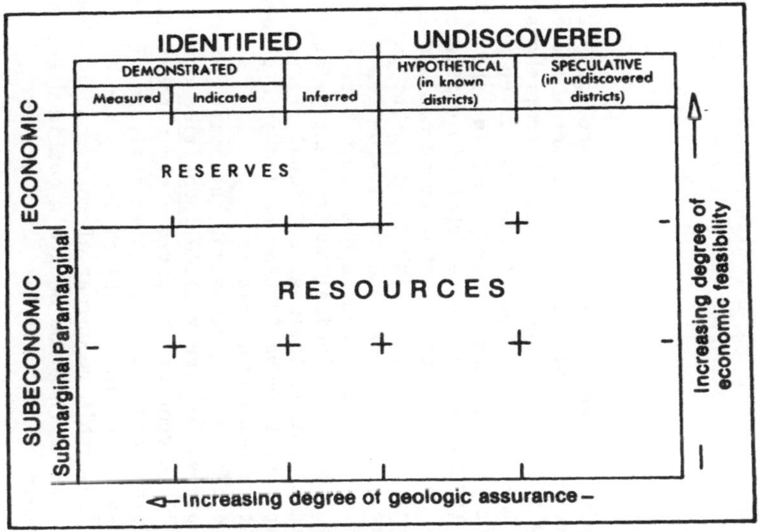

Figure 1.1 Resource classification diagram by the U.S. Geological Survey and the U. S. Bureau of Mines.

Source. U. S. Bureau of Mines and U. S. Geological Survey, 1976.

Naturally-occurring coking coal is bituminous; however, only a small percentage of bituminous coal has the necessary constituents to be a coking coal. Thus, coke is usually manufactured from blends of two or more coals of different rank and composition. In the process, these coals are subjected to a carbonization process in a high temperature oven and then rapidly cooled in water. When cool, coke is crushed and sized for use in a steel blast furnace. About a ton of coke is used to produce a ton of steel, and production of a ton of coke requires 1.5 tons of coal for the blending process. The process will also yield about one-half ton of coke-oven gas, oils and tars.

AMERICA'S COAL DEPOSITS

The basic geological knowledge collected and analyzed up to the date of a resource report usually influences the estimated size of coal resources. The most recent comprehensive report to date on United States coal resources has been compiled by Averitt (1975). And in his report for the U.S. Geological Survey (USGS), Averitt noted that his estimate of 3,968 billion tons for United States coal resources was 23 percent higher than any previously published data. The pattern was similar for previous estimates and the lesson seems clear; the estimates of a particular resource base vary directly with increased knowledge of the geological environment related to that resource. Consequently, the next USGS coal resource estimate can be expected to have further refinements as new geological information becomes available.

Based on current resource estimates, the United States contains about 23 percent of the world's identified coal resources and about 20 percent of the world's estimated total coal resources. On the basis of a weighted analysis of data on resources of fossil fuel in the United States (as of January 1, 1974) recoverable resources of coal contain about ten times more heat value than the nation's combined recoverable resources of petroleum and natural gas (Averitt, 1975, pp. 1-2).

Coal resources are distributed somewhat irregularly across the United States, but the resources can be grouped, generally, into the major coal fields displayed in Figures 1.2 and 1.3. And to place Colorado coal resources and production in some perspective, Table 1.2 illustrates the coal reserve base of the United States as of January 1, 1974, by state and by method of mining. Most of this coal is available to be mined, but mining techniques vary in efficiency. For example, recovery of coal by underground mining is about 50 percent efficient; surface or strip mining recovery averages 80 percent; and auger mining can reach up to 55 percent efficiency, but averages about 35 percent or less. It follows, therefore, that estimates of reserves and resources should be reduced significantly because of the limitations of present mining

Figure 1.2 Coal fields of the conterminous United States
Source: Averitt, 1975, p. 5.

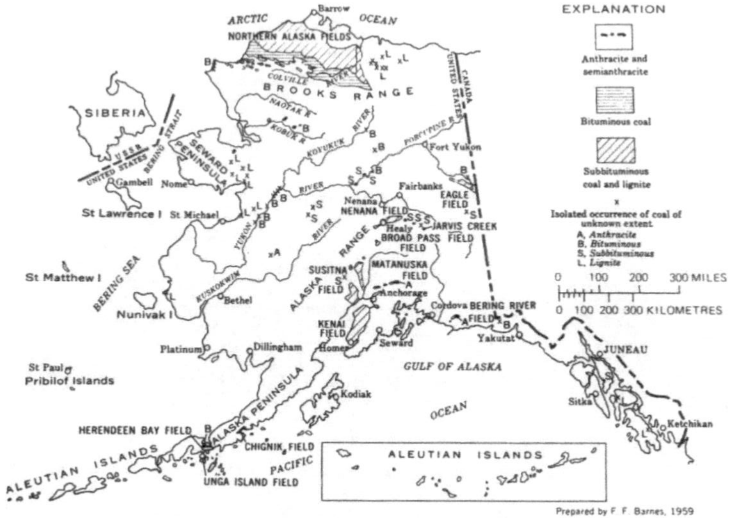

Figure 1.3 Coal Fields of Alaska.
Source: Averitt, 1975, p. 6.

Table 1.2

COAL RESERVE BASE OF THE UNITED STATES, JANUARY 1, 1974 AND BY METHOD OF MINING (Data from U.S. Bureau of Mines (1974b, p. 4) in millions (10^6) of short tons. Figures are for coal in the ground.)

	Potential Mining Method		
State	Underground	Surface	Total
Alabama	1,798	1,181	2,982
Alaska	4,246	7,399	11,645
Arizona	(1)	350	350
Arkansas	402	263	665
Colorado	14,000	870	14,870
Georgia	1	.....	1
Illinois	53,442	12,223	65,665
Indiana	8,949	1,671	10,623
Iowa	2,885	(1)	2,885
Kansas	(1)	1,388	1,388
Kentucky, eastern	9,467	3,450	12,917
Kentucky, western	8,720	3,904	12,624
Maryland	902	146	1,048
Michigan	118	1	119
Missouri	6,074	3,414	9,488
Montana	65,165	42,562	107,727
New Mexico	2,136	2,258	4,394
North Carolina	31	(2)	31
North Dakota	.....	16,003	16,003
Ohio	17,423	3,654	21,077
Oklahoma	860	434	1,294
Oregon	1	(2)	1
Pennsylvania	29,819	1,181	31,000
South Dakota	.....	428	428
Tennessee	667	320	987
Texas	.....	3,272	3,272
Utah	3,780	262	4,042
Virginia	2,971	679	3,650
Washington	1,446	508	1,951
West Virginia	31,378	5,212	39,590
Wyoming	27,554	23,671	51,118
TOTAL	297,235	136,713	433,948

Source: Averitt, 1975, p. 33.

(1) Data insufficient to establish reserve base
(2) Less than 1 million tons

technology (Wang, 1978).

Assuming that about 50 percent of the underground reserve base (297 billion tons) and 80 percent of the surface reserve base (137 billion tons) may be ultimately recovered, the United States will have available 258 billion tons of coal for resource development. If the 1977 coal production rate of 688 million tons were to continue at the same pace, then the United States would have a coal supply for the next 375 years. However, a growth factor of 3 percent per year, added to the 1977 base-year production, reduces the supply estimate to 84 years. Even a growth factor of 5 percent per year (perhaps unattainable under any circumstances) would still provide a coal supply for the next 60 years (Stanwood, 1978a). Theoretically then, our coal reserves are large enough to satisfy a large portion of the nation's future energy needs--perhaps until alternative energy systems are developed and are on-line.

In the West, coal production has increased substantially in the past several years because: (1) shortages and high costs of alternative fuels are forcing many powerplants across the country to convert to coal; (2) air quality concerns are causing coal consumers to turn from their traditional sources with high sulfur content to the low sulfur coals found in the West; and (3) lower production costs of strip-mining coal in the West, due to much thicker coal seams, make it economically competitive (Lowrie, 1977, p. 175). If this trend continues, the western states will play a major role in future coal production for the United States. This may well happen even though federal air pollution control regulations requiring the use of "best available technology" to control power plant emissions may reduce the economic benefits of burning low-sulfur western coals.

COLORADO'S COAL DEPOSITS

Colorado has an estimated bituminous and sub-bituminous coal reserve base of about 14.9 billion tons (see Table 1.2). About 94 percent of this reserve base is minable only by underground methods. Like many western coal resources, those in Colorado often occur in relatively thick (3-20 feet or more), closely spaced, multiple beds. The beds dip steeply in places and are usually highly lenticular, variable in thickness, and overlain by roof-rock of variable character. The thicknesses of the overburden change rapidly. The net result of these variable factors may greatly lower the percentage of recoverable coal (perhaps to as low as 15-20 percent of the in-place reserve base) (Murray, 1977, p. 2).

Colorado's coal resources are divided into eight coal regions and twenty coal fields, as displayed in Figure 1.4. These eight regions are thought to contain over 454 billion tons of total, in-place coal resources, some to depths of 6,000 feet (Averitt, 1975, p. 14). This amounts to almost 11 percent of the total coal resources of the United States

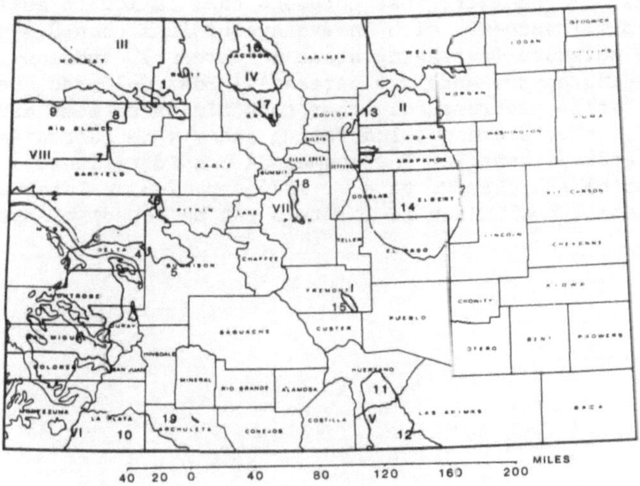

COAL FIELDS

I	Canon City	1.	Yampa	11.	Walsenburg
II	Denver Basin	2.	Book Cliffs	12.	Trinidad
III	Green River	3.	Grand Mesa	13.	Boulder-Weld
IV	North Park	4.	Somerset	14.	Colorado Springs
V	Raton Basin	5.	Crested Butte	15.	Canon City
VI	San Juan	6.	Carbondale	16.	North Park
VII	South Park	7.	Grand Hogback	17.	Middle Park
VIII	Unita	8.	Danforth Hills	18.	South Park
		9.	Lower White River	19.	Pagosa Springs
		10.	Durango	20.	Nucla-Naturita

Figure 1.4 Summary of coal resources in Colorado.
Coal Regions and Fields in Colorado

Source: Hornbaker, 1976, p. 1.

and ranks Colorado fourth in national coal resources. Additionally, Colorado ranks first in identified low-sulfur (less than 1 percent) bituminous coal resources in the United States (Murray, 1978, p. 1). This low-sulfur coal blends well for combustion with the high-sulfur coal of other states. Table 1.3 illustrates Colorado's identified resources as of January 1, 1974.

Coal in Colorado ranges from low grade lignite (in the Denver Basin), through sub-bituminous and bituminous, to anthracite (found in the Crested Butte area in Gunnison County). In quality, most Colorado coal is low in sulfur (0.2 to 1.1 percent, with an average of less than 0.5 percent), low in moisture (averaging about 12 percent), and low in ash (averaging approximately 6 percent). Most Colorado coals require little pretreatment other than sizing to meet market demands; they are nonagglomerating and may be carbonized in fluidized systems (See Environmental Considerations, Chapter 6) (Hornbaker, et al., 1976, pp. 2-3). Information on existing coal mines in Colorado can be found in Appendix A.

Table 1.3

IDENTIFIED COLORADO COAL RESOURCES (January 1, 1974)
(Billions of Tons)

Coal	Est. Original[a] Identified	Depletion to 1/1/74 Production	Depletion to 1/1/74 Mining Losses	Remaining 1/1/74
Anthracite[b]	0.090	0.006	0.006	0.078
Bituminous	110.0	0.443	0.443	109.1
Subbituminous	20.0	0.216	0.126	19.7
Lignite	20.0[c]	.....	.....	20.0
Total Resources				148.9

Source: Averitt, 1975, pp. 11-15.

Note: Hypothetical resources: (geologically predictable, but may or may not be recoverable economically)
1. Unmapped or unexplored areas (0-3,000 ft overburden) = 146 billion tons.
2. (3,000-6,000 ft) = 145 billion tons.
3. Total hypothetical resources = 291 billion tons. (Cf. Figure 1.1)

(a) Estimates include beds of anthracite and bituminous coal about 14 in. or more thick; subbituminous coal and lignite have beds about 2.5 ft thick.
(b) Includes anthracite and semianthracite.
(c) The total is the corrected number supplied by K. Murray, Colorado Geological Survey. That reported in Ref. 1 is in error.

2
Exploration Overview

The search for coal poses many of the same problems and uses many of the same techniques as those involved in exploring for other fossil fuels, metals and industrial minerals. Finding coal beds which can be recovered through surface mining differs only slightly from exploring for deeper coal. And since much is already known about the location of coal fields, further exploring concentrates on choosing locations where the coal may be mined most efficiently.

Coal with almost no exceptions, is found as strata, or bands, in sedimentary rock. Deposits are explored to determine: (1) extent (areal coverage, bed thickness); (2) quality (specific gravity, sulfur content); (3) rank (percentage of fixed carbon, heat content--anthracite, bituminous, etc.); (4) depth of deposit (to determine the economics of underground or surface mining); and (5) potential environmental impacts of recovery.

In general, exploration data are collected initially by mapping outcrops of coal beds. This is followed by core drilling to test the thickness and continuity of the bed. Table 2.1 identifies the large number of parameters which are identified in the exploration phase and which are essential for determining if the project is to proceed. It should be noted that, in addition to the more technical aspects of exploration, environmental and socio-economic impacts also require evaluation. When the geology is suitable but outcropping only meager, more sophisticated approaches to coal exploration may be used.

In the western United States, coal deposits at strippable depths are often thick, low grade and severely split, so their thicknesses may be variable within short distances. Research done by the USGS and the Colorado School of Mines has identified geophysical approaches to augment geologic and borehole data, and at the same time reduce the extent of drilling needed to delineate a deposit and its attendant cost. These approaches include high-precision gravity surveys to locate

Table 2.1

FACTORS CONSIDERED IN COAL EXPLORATION

Project Information Obtained from Coal Coring	Coal Exploration	Exploration Program Essential Input	Additional Data
Extent of Recoverable Reserve	Reconnaissance	Geology (Discovery)	Presence or absence of aquifers
Coal Quality	Initial Leasing	Mining Engineering	Competency of overburden strata
Type of Beneficiation Needed	Drilling or Geologic Mapping	Geographic Location (Transportation)	Presence or absence of toxic substances
Geologic Structure	Leasing of Total Project	Quality (Marketing)	Type of floor or substrata
Unusual Conditions	Development Drilling		
Type of Overburden	Feasibility		
Anticipated Rate of Methane Emission			

Sources: Adams, 1977; A. Lees, 1978.

discontinuities; magnetic methods to map burn facies; seismic seam waves to identify seam boundaries; and a combination of borehole logging, seismic seam wave and seismic reflection techniques for precise mapping (Hasbrouck and Hadsell, 1977, pp. 187-197).

3
Recovery Methods

Much of the western coal reserves are sufficiently low in sulfur content to meet power plant air emission standards when burned; however, most of this coal can only be obtained by underground mining. The special characteristics of western coalfields--unusually thick coal seams, closely spaced multiple seams, susceptibility of some coals to spontaneous combustion, and the presence of massive sandstone beds above and below the coal seams--require some adaptation of "eastern" mining methods (U.S. Bureau of Mines, 1977).

The decision to use surface or underground mining is largely dependent upon such factors as depth of overburden, seam thickness, deposit size and local geology. With improvements in equipment and technology, the trend has been toward an increase in surface mining. Hence, more than 50 percent of U.S. coal is now recovered by this method. The significance of changes in earth moving techniques illustrating this trend may be quantified by considering the changes in overburden-to-seam-thickness ratios which can be economically mined. They have increased from 10:1 (1965) to approximately 30:1 today (University of Oklahoma, 1975).

The discussion that follows reviews present surface and underground mining methods and identifies changes that may be applied to western coal.

SURFACE MINING

Surface mining, often called "strip mining," requires removal of the overburden to allow recovery of the mineral directly from the surface. The two major types of strip mining are contour and area mining. Contour mining is used in hilly or mountainous areas where the overburden is removed along an outcrop so that a flat surface, or bench, is formed. Coal can then be mined from the exposed surface and from the highwall by using large drills or augers.

"Area mining" recovers coal from flat or rolling terrain. The process involves a trench excavation starting on one side of the field to expose the seam. As the coal is removed, the trench is filled with fresh overburden stripped from the adjacent cut as the operation advances across the seam.

Mining operations associated with surface mining include the following steps.

1. Surface preparation - Vegetation is removed from the mine site. When it is sparse, it can be removed with a dragline along with the overburden. This preparation phase also includes the construction of access roads, maintenance and personnel facilities, and installation of necessary utilities. Bulldozers, scrapers and loaders are used in preparing the surface. Topsoil is usually transported to a stockpile area for later use in surface reclamation.

2. Fracturing - This step involves the drilling of blast holes, along with the actual blasting, to fracture the overburden and the coal seam thereby making the seam easily accessible to the mining equipment.

The equipment selected for use depends upon its flexibility in relation to the nature and quantity of the material to be moved, the distances involved and the available surfaces for material transport. Typically, mining operations use some combination of the following: (Bertoldi, 1977)

a. Small, mobile tractors. (Including bulldozers, scrapers, and front-end loaders). These are versatile, easily maneuvered, able to negotiate steep grades, and able to dig and transport their own load.

b. Shovel-truck systems. Diesel or electrically powered units, having capacities of up to 130 cubic yards, can traverse a bench and scoop fractured overburden or coal into their huge buckets. These systems offer great versatility in handling material. Maximum production, however, requires rigidly engineered, coordinated and scheduled operations.

c. Draglines. These electrically powered machines have a very large capacity for a "single bite" in excavating. They actually perform two functions: excavating and conveying. With bucket capacities ranging from 30 to 220 cubic yards, the draglines move along the bench and place their buckets on the overburden to be removed. The bucket is loaded by dragging it towards the machine, then lifting, rotating and dumping it onto the designated location. Draglines are also adaptable to pitching or dipping coal seams.

d. Bucket wheel excavators. Only larger mines with soft overburden can justify the cost of using bucket wheel excavators. The unit contains a bucket wheel (of up to 50 feet

Plate 3.1 "Effie", Energy Fuels Corporation's 55-cubic-yard dragline, is shown here removing overburden to expose coal seams at the Company's surface mine 25 miles southwest of Steamboat Springs. This dragline is the largest in Colorado. (Courtesy: Energy Fuels Corporation)

or more in diameter) mounted on a boom (of up to 400 feet long). By rotating the wheel, the buckets facing the cut are loaded and the material is then emptied onto a conveyor which carries it to the in-mine transportation system. The use of bucket wheel excavators can reduce the number of levels in the mine but is limited to deep surface mines with an overburden of several hundred feet.

Generally, the unit cost of moving a cubic yard of material by dragline is the least expensive of the options; however, each mining situation is unique and must be evaluated accordingly.

UNDERGROUND MINING

It is estimated that underground mining will produce 400 million tons of coal nationally by 1985 (about one-third of U.S. projection at that time). These projections indicate a growth of 39 percent over current production levels or an annual growth rate of slightly over 3 percent. These limited growth projections are based upon an estimated tripling of western coal output, with an accompanying projection of increases in surface mining (Hunter, 1977). The basic underground mining methods used are: (1) continuous mining machines in room and pillar operations; (2) longwall and the newly developing shortwall methods (which can also use continuous miners); and (3) cutting machines or auger mining. Table 3.1 shows the production for each system for the period 1970-1975.

Longwalling, shown at less than 4 percent of national production in 1975, is expected to grow to 15 percent by 1985 and possibly to account for as much as 60 percent of underground coal mining by the end of the century (Merritt and Davis, 1977). Presently, 25 percent of western mines use longwalling, which may produce up to ten times more coal per miner than other methods (Mining Record, 1978).

In each type of underground mining, the coal seam is reached by digging or boring either a vertical shaft, a horizontal portal, or a slanting tunnel. Before this, however, basically the same procedures are used for preparing the surface as those used in surface mining. However, major differences in mining techniques begin once the deposit is reached. Underground mining includes a number of techniques.

1. Room and pillar mining is started by excavating a passageway through the coal seam. By mining coal in directions away from this passageway, rooms are formed, leaving sections of the seam remaining as pillars to support the overlying rock strata. Factors determining how large a room can be mined are related directly to the nature of the surrounding strata. These factors include the bedding strength of the coal itself as well as the types of materials overlying and underlying the seam. In addition to the coal pillars, mechanical supports such as roof bolts are used.

Table 3.1

UNDERGROUND COAL PRODUCTION BY EXTRACTION METHOD, 1970-1975, MILLIONS OF TONS

	Hand Loading		Continuous Mining Machines		Longwall		Cutting Machines		Total Underground
	Total	%	Total	%	Total	%	Total	%	
1975	4.8	1.7	178.6	61.6	10.4	3.6	96.0	33.1	289.8
1974	4.9	1.8	171.3	61.8	9.6	3.4	91.5	33.0	277.3
1973	4.3	1.4	178.6	29.7	9.4	3.1	107.3	35.8	299.3
1972	4.9	1.6	178.4	58.6	7.8	2.6	113.0	37.2	304.1
1971	4.7	1.7	152.9	55.4	6.5	2.4	111.7	40.5	275.8
1970	5.6	1.7	169.9	50.1	7.1	2.1	156.2	46.1	338.8

Source: Merritt and Davis, 1977

Typical eastern underground mines, where coal seams vary from 2-6 feet in thickness and where the coal and surrounding materials are of low strength, use long (several hundred feet) and narrow (10-20 feet wide) rooms. This arrangement permits the use of specialized mining equipment for removing the face of the seam and loading it onto some type of conveyance. Most U.S. operations use either conventional mining (which involves machine undercutting, blasting, and mechanical loading of the coal), or continuous mining (in which one mechanical device performs undercutting and loading operations).

2. Longwall mining involves removing large quantities of coal from a seam face which may be 600 feet long in a single, continuous operation. Shortwall mining, a variation of the above method, refers to seam faces of some 150 feet in length. In longwall operations shearing machines, such as plows or planers and shearing drums, are used; in shortwall mining, the "continuous miner" is used. In this case, the shearer moves in both directions across the working face of the seam between two access passageways (or galleries), dropping coal onto a conveyor which carries it to the main mine transportation system. Hydraulic jacks support the roof over the working face; they are advanced towards the new face as the coal is removed. And as the jacks are moved forward the roof is allowed to collapse behind.

Longwall mining is presently used more in the eastern mines, but is spreading toward the west. Additional techniques now under development are designed to address the thicker and more steeply pitching coal seams found in the west. A surge in Western longwall mining is expected by the early 1980's (Merritt and Davis, 1977).

3. Auger Mining is often used in surface mining to recover coal from exposed seams in the high wall. If a large subsurface seam remains after surface operations are completed, either an underground mine or recovery through augering can be attempted. It is possible to use augering if the tonnage is small or the seam is too thin. Often augers are used to mine coal which is physically or economically impractical to recover by other means.

4. Advanced Methods. Increased federal funding, largely to private companies, is helping to develop new concepts in mining equipment and technology. The aim is to increase production efficiency and percentage of coal recovered in safer, healthier mines (Cheronis, 1977). One of these advanced methods combines some of the better features of longwall mining with those of auger mining and is expected to provide a significant improvement in production (as much as 6000 tons per 8 hour shift) over existing systems. This method appears to offer a higher ratio of coal recovery, increased safety and reduced leadtime in bringing a new mine to

Plate 3.2 Small utility LHD (load-haul-dump) or end loader conterpart being used in the installation of vent tubing. This is a diesel powered, articulated loader used for many utility operations, including clean-up. (Courtesy Colorado Westmoreland Coal Co.)

full production.

Another new technique, borehole mining, is concerned with the problems of mining steeply pitched coal seams; it is an attractive possibility for some areas in the West. Hydraulic mining, where water is available, is being investigated for cutting coal in an economical, safe and environmentally acceptable manner. Through his work at the Colorado School of Mines, Dr. F. D. Wang has become one of the national leaders in this technology.

The recovery efficiencies for the various types of underground mining are estimated as: room and pillar, 45-50 percent; longwall mining, 65-75 percent (and possibly as high as 80-90 percent); and auger mining, about 35 percent (Wang, 1978).

4
Coal Cleaning

As noted earlier (see Nature of the Resource, Chapter 1), coal is a heterogeneous material comprised of organic, or combustible matter, and mineral matter. The mineral matter in coal contains noncombustible impurities which are broadly divided into ash-forming and sulfur-forming (or contributing) products. Each of these categories is further divided into two groups: those impurities chemically part of the coal which are not removable by mechanical means, and those not chemically bound to the coal which are removable by mechanical processes. Physical coal cleaning, which includes beneficiation (preparation by washing), is used only in the second category of impurities (U.S. Department of the Interior/Environmental Protection Agency, 1977).

In order to meet New Source Performance Standards, which allow a maximum release of 1.2 lbs. of sulfur dioxide (SO_2) per million Btus of coal burned, mixing coals of various sulfur content is often required. In fact, a recently developed computer simulation model based on operating experiences of the Pennsylvania Electric Corporation analyzes the extent of cleaning and preparation needed for various coals (Jacobsen, 1978).

The first step in physical cleaning is crushing the coal to allow mechanical removal of impurities. One of the most common processes makes use of the differences between the specific gravity of coal (about 1.3) and that of the impurities (which is greater than 1.3). Slurrying coal in water allows the impurities to settle. Equipment now in use slurries between 500-1000 tons of coal per hour. The amount of crushing, screening and washing needed is described in Jacobsen's 1978 model.

Since the lower heat value impurities have been removed in cleaning, the Btu content per unit weight of cleaned coal is increased. An added benefit is that transportation costs will be less per Btu delivered since the Btu value per unit of weight increases. However, the total Btu content is reduced from that of the raw coal because some coal is lost in the process.

Besides cleaning coal prior to burning, other options are being explored. Cleaning during burning (called fluidized bed combustion) is a technology now under research and development. Post-combustion devices used to clean coal are called flue gas desulfurizers (FGDs) (See Technology Factors, Chapter 8). Combinations of coal cleaning and stack gas scrubbers, or FGDs, are amenable to technological and economic trade-offs as shown in Figure 4.1. For example, coal cleaned prior or during combustion will reduce the quantities of flue contaminants needing treatment. Further, burning clean coal significantly reduces the sludge disposal problem from FGDs.

Cost savings from using less expensive FGD systems must be balanced against the additional costs for the cleaning operation. Assessing the benefits of the systems, singly or in combination, must rely on results of the experience now being compiled. Table 4.1 compares cost data for combinations of physical coal cleaning with FGDs and for FGDs alone. The potential environmental degradation from FGD sludge disposal should also be further evaluated (See Environmental chapter for more information on FGDs).

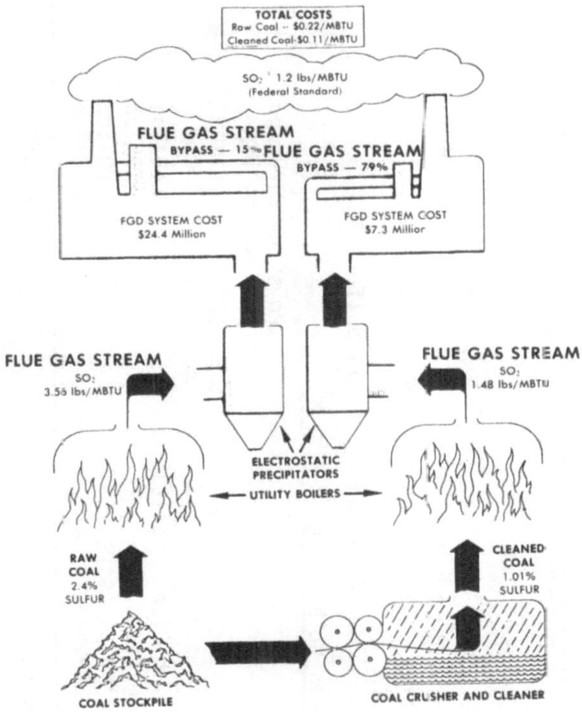

Figure 4.1 Physical coal cleaning with flue gas desulfurization vs. flue gas desulfurization alone.

Source: U.S. Department of Interior and Environmental Protection Agency, 1977.

Table 4.1

CASE STUDY - COAL CLEANING & FGD VERSUS FGD ALONE (To determine most cost effective approach for a new coal-burning utility plant to meet emission standard of 1.2 lbs. SO_2 per million Btu (MBTU)

Case Conditions:

Coal Use Area:	Tonawanda (Buffalo), New York	
Coal Source Area:	Cambria County, Pennsylvania Coalbed: Lower Freeport	
Raw Coal Characteristics:	11.4% Ash	2.4% Sulfur
Cleaned Coal Characteristics:	6.7% Ash	1.01% Sulfur

Cost of Alternate Approaches to Meeting Standards:

	Physical Cleaning Followed by FGD	FGD Alone
Coal Cleaning Cost		
Amortization of Cleaning Plant, Capital Cost	$ 0.68/Ton	$ -0-
O & M Cost of Cleaning Plant	0.75/Ton	-0-
Cost of Coal Lost During Cleaning	1.56 To 2.00/Ton	-0-
Taxes & Insurance of Cleaning Plant	0.12/Ton	-0-
Total Cleaning Cost	$3.11 To 3.55/Ton	$ -0-

Cost of Flue Gas Desulfurization (FGD)		
Amortization of FGD Capital Cost	$ 0.92/Ton	$ 2.92/Ton
Fuel & Electricity of FGD System	0.17/Ton	0.66/Ton
O & M Cost of FGD System	0.72/Ton	2.24/Ton
Total FGD Cost	$ 1.81/Ton*	$ 5.82/Ton*
Benefits of Using Cleaned Coal		
Increased Heat Content	$1.21 To 1.40/Ton	$ -0-
Transportation Savings	0.30/Ton	-0-
Ash Disposal Savings	0.21/Ton	-0-
Pulverizing Savings	0.02/Ton	-0-
Maintenance & Other Savings	0.23/Ton	-0-
Total Benefit of Cleaning	$1.97 To 2.16/Ton	$ -0-
Net Cost (Costs Less Benefits)	$2.76 To 3.39/Ton	$ 5.82/Ton
Converted to per MBTU	$0.10 To 0.12	$ 0.22

SOLUTION:

The combined approach is the most cost-effective, compared to a 100% additional cost for meeting the standard by FGD alone.

Source: U.S. Department of Interior/Environmental Protection Agency, 1977.

*Significant difference in cost of FGD results from 21% of the flue gas being processed versus 85% in the more expensive approach to meet the required emission standard.

5 Commercial Development Considerations

LEASING OF COAL LANDS

Federal, State, and county governments, as well as private parties, hold leases on coal lands. Table 5.1 shows the distribution of leases by source in the sixty-one licensed coal mining operations registered in Colorado as of April 1978. The private lease is the most prevalent and, historically, provides for most of Colorado coal production. However, federal leases are playing an increasingly important role in production.

Table 5.1

COLORADO COAL LEASE SOURCES (operating mines) AS OF APRIL 1978

Lease Type	Number of Mines
Private only	40
Federal only	9
State only	1
Private and Federal	5
Private and State	4
Private and County	1
Federal, State and County	1

Source: Colorado Division of Mines, 1978, pp. 1-2.

The need for federal leases is increasing because these leases form a logical mining unit (LMU) along with private, state, and other federal leases. An LMU is a mining unit in which the coal reserve can be mined efficiently and economically, usually by one operator. Today there are many existing mining operations in Colorado which need a tract of federal coal land to complete an LMU, yet this additional land has been very difficult to obtain.

Currently, there are 115 federal leases in Colorado covering an area of almost 122,000 acres. Included in these totals are two recent short-term leases: one to Colorado Westmoreland (311 acres) in March of 1978 and one to Energy Fuels Corporation (263 acres) in June of 1978. The debate among the various interest groups on lease acquisition today centers around federal leasing policy, which has gone through numerous changes in the past ten years. Martin (1978, pp. 1-4) has developed a brief history of federal coal leasing which is presented here.

Prior to 1971, there was significant activity in federal coal leasing, much of it for speculation purposes. At the same time, there were significant decreases in coal production on federal lands. This discrepancy between the amount of coal leased and the amount being produced was the primary reason the Department of the Interior (DOI) declared a moratorium on federal coal leasing in 1971. But, perhaps as important was the growing belief that there should be a coal leasing system which recognized alternative and potential uses of public lands (e.g., recreation, wildlife habitat and livestock grazing) in addition to coal development.

In February of 1973, DOI initiated short term leases to allow existing coal operations to continue mining on adjacent federal lands during the moratorium period. The criteria for short term leasing changed three times after they were first developed. Then, in October of 1973, DOI initiated a new federal leasing system called the Energy Minerals Allocation Recommendation System (EMARS). A draft Environmental Impact Statement followed in May 1974 assessing this new policy. A final Environmental Impact Statement later released in September 1975 was significantly changed. The acronym EMARS now stood for Energy Minerals Activity Recommendation System and the system now required industry to nominate the coal areas to be leased. After some review, EMARS was put into practice in January of 1976.

In August 1976, Congress passed the Federal Coal Leasing Amendments Act of 1975. Besides abolishing noncompetitive coal leasing, this legislation required comprehensive land use planning prior to leasing federal lands, and ruled that social, economic, and agricultural impacts be considered in leasing decisions. While this Congressional mandate did not conflict with the on-going EMARS process, criticism had already been mounting over the complexity of EMARS requirements. Finally, a U.S. District Court decision, National Resources Defense

Council (NRDC) vs. Hughes (DOI) held that the EMARS program was not appropriate since the DOI programmatic EIS on federal coal leasing failed to adequately consider the alternative of "no leasing" on federal lands. As it turned out, no coal leases ever completed the EMARS process.

The major outcome of the NRDC vs. Hughes case has been to prohibit DOI from any further coal leasing except for short term leases. In addition, the short term leasing criteria are more stringent; short-term leases can extend coal supplies for only three years at the current production rate and are granted only to fulfill existing contracts (which has been interpreted to mean contracts existing prior to the Court's decision). The DOI is required to write a new programmatic EIS, do additional studies on existing leases, and design an acceptable management system before federal coal leasing can resume. Production forecasts derived from projected demand for coal in various regions in the United States were recently compiled by the Department of Energy (1978a), and these forecasts will be used to help determine the future need for coal leasing.

The Colorado Department of Natural Resources (1978, p. 3), as official representative of the State in matters associated with federal coal leasing, would like the following issues thoroughly considered and evaluated in any new federal leasing program:

1. State participation in federal land use planning.
2. Recognition of the uniqueness of issues and problems of individual states.
3. National planning.
4. Regional planning.
5. Regional environmental impact statements.
6. Coal development priority areas or zones.
7. Social and economic impact assessment.
8. Short term leases and preference right (noncompetitive) leases.

The preferred federal coal leasing and management program, released in a draft Environmental Impact Statement by the Bureau of Land Management in December 1978, includes eight major elements:

1. A planning system, involving close consultation with state and local governments, industry, and the public, aimed at helping (1) to decide which areas of federal coal reserves would be considered acceptable locations for coal production, and (2) to delineate, rank, and select for sale specific tracts of coal.
2. A system for evaluating, in conjunction with the Department of Energy, the national demand for coal and for determining production which should be stimulated by the leasing of federal coal.

3. Procedures for conducting sales and issuing leases.
4. Post-lease enforcement of terms and conditions.
5. Procedures for management of existing leases issued prior to implementation of the new program.
6. Procedures for processing existing preference right lease applications.
7. A strategy to integrate the environmental analysis requirements of the National Environmental Policy Act of 1969 in the new program.
8. Procedures to start-up the new program and to offer lease sales in emergency situations.

Six additional alternatives were considered in this draft programmatic EIS, and it is expected that a decision on the new program can be reached by mid-1979, so that leasing could resume in mid-1980.

If the proposed federal leasing system is approved, the regional EIS concept will be an integral part of the federal leasing program. The preparation of regional EISs should eliminate the need for site-specific EISs in most cases if the concept evolves successfully. The regional EIS concept consists of a regional analysis along with site-specific analyses considering the impacts of proposed new developments in the study area. If the EIS is then approved in this regional format, the companies involved may avoid possible delay and expense because individual site-specific EISs will not be necessary.

An example of the regional EIS concept is the Northwest Colorado Coal Regional EIS, released in final form in early 1977, just after the EMARS program was announced. Since the EIS did not include discussions and impacts of EMARS, it was judged to be inadequate. A supplemental statement was to be prepared to deal with this inadequacy, however, when the federal leasing policy was thrown into limbo by the <u>NRDC</u> <u>vs.</u> <u>Hughes</u> decision, the thrust of the study (released in late 1978) was changed to include an update of environmental conditions in the area and the appropriateness of new regulations.

In the case of leasing State coal land, except where the Federal Government owns the mineral rights, State lands are leased through oral bids by the State Board of Land Commissioners. Coal leases on private lands, except those on which the Federal Government or the state owns the mineral rights, are usually obtained through negotiation with the owner. Information on ownership of specific tracts of land or on mineral rights is available only from the county assessor of the county in which the land is located (Speltz, 1976, pp. 11-12).

PERMIT SYSTEM

The number of permits, licenses, and agencies (both federal and State), involved with the mining operations varies with each specific site. Generally, a wide variety of federal and State regulatory agencies may require a combined total of 15-20 permits before mining can actually begin. As a normal procedure, an operator will not submit all of the permit applications required for mining until he is certain he will be able to market the coal; in addition, he will need to assess mining production rates, delivery time and other economic factors before making some permit applications. It is apparent that timing of applications is very important (Jones, 1977, p. 136).

Quite often the permit process required of the coal mine operator today is time consuming and confusing. In late 1976 the Mountain Plains Federal Regional Council (1977, pp. 1-11 and 4 flowchart plates) through its Energy Regulatory Review Committee, began a study to unravel this complicated process. The following information is taken from that Committee's findings.

The study lists the existing Federal Government procedures that must be followed before beginning construction of a surface coal mine on federal lands and/or coal deposits. In the process, the study makes recommendations on how the procedure can be improved without sacrificing any environmental and safety performance standards or diminishing the public's interest in its ownership of the coal resource. Certain minor inaccuracies already exist in the flowcharts, but it is anticipated that they will be periodically corrected and updated in the future. Although the leasing section of the report is no longer current and inaccuracies exist in the permit section of the flowchart, the document contains much relevant information. The complexity of the permitting process is clear and the corrected and updated version planned by the Committee is certainly needed.

The study does not cover requirements for Indian reservations, private or State-owned property, or State and local governments. However, two states (Colorado and Wyoming) have worked closely with the Committee and are developing documents describing their state requirements for opening a surface coal mine. In addition, the Colorado Energy Research Institute is following up on this work by preparing a flowchart which shows the State permitting process. Additional work needs to be done to determine how federal and State processes and requirements could be improved to eliminate overlaps and how they may be effectively linked together.

During the research, it became clear that no single agency, company, or entity understood the total permitting process. Each, in fact, focused on only limited elements of it. A comprehensive analysis of the total permitting process should greatly assist government agencies, private companies and public interest groups to understand and improve the system.

To some extent twelve federal agencies are involved in issuing necessary licenses, permits and approvals before a surface coal mine can begin construction. Most important among these, in terms of the amount of time required to obtain the necessary permits, are the Bureau of Land Management (BLM) and Geological Survey (USGS) within the Interior Department, as well as the Interior Department itself.

Other federal agencies having requirements (which can generally be met at the same time as those of BLM and USGS) are the Forest Service, Environmental Protection Agency, Fish and Wildlife Service, Federal Communications Commission, Federal Aviation Administration, Internal Revenue Service, Corps of Engineers, Interstate Commerce Commission, Mine Safety and Health Administration, The Bureau of Alcohol, Tobacco and Firearms in the Treasury Department, and most recently, the Office of Surface Mining.

Table 5.2 lists estimates of elapsed time from the date when permitting, lease acquisition and mine plan development begin to the date when mine construction is authorized. Depending on the company's situation, total elapsed time ranges from 1.0 to 11.8 years. It can be inferred from looking at case 2 (company has lease, explores and develops mining plan) that the permitting process during exploration and mine plan development takes 7.1 years. Much of this time may be spent merely waiting for various permit approvals, rather than actually exploring or planning.

The elapsed time of some of these cases includes the period to acquire a lease under obsolete leasing processes. However, it may be safe to assume any new leasing system timetable will probably not be radically different from those used by the Energy Regulatory Review Committee.

To explore this issue further, an analysis was made for each of the eight cases to determine the amount of elapsed time that might be attributed to administration, environmental analysis, exploration, etc. Administrative time, it was discovered, ranged from 39-86 percent of the total processing time, indicating that administrative activities might well be a fertile area for developing more efficiency in the process.

In reference to the State regulatory process, the Colorado Environmental Permit Directory helps to identify State permits required for development. The Directory (Kinney, 1977, p. 2) reports that a mining operation must obtain State permits from two basic sources: the Department of Health and the Department of Natural Resources. Within the Health Department, the Air Pollution Control Division, Water Quality Control Division (See Environmental Legislation chapter), and the Water Quality Engineering Division require various permits; the Department of Natural Resources requires permits from the Division of Mines (concerning safety and licensing) and the Mined Land Reclamation Board (concerning mining and environmental plan approval). Specific information on these State permits can be found in the Directory.

Table 5.2

ELAPSED TIME FOR SURFACE COAL MINE AUTHORIZATION

Case	Description	Years
1	Company has lease and develops mining plan	3.4
2	Company has lease, explores and develops mining plan	7.1
3	Company waits for BLM list of tentative lease tracts, bids, obtained lease, explores & develops mining plan	7.4
4	Lease modification (assumes new exploration or mining plan would not be required)	1.0
5	Company obtains preference right lease, explores & develops mining plan	8.7
6	Company obtains lease under short-term criteria and modifies existing mining plan	1.2
7	Company nominates coal tract, obtains competitive lease, explores and develops mining plan	9.5
8	Company conducts preliminary exploration, nominates tract, obtains competitive lease, explores and develops mining plan	11.8

Source: Mountain Plains Federal Regional Council, 1977, p. 4.

CAPITAL AND OPERATING COSTS

Capital investment estimates vary for surface and underground mines. Since numerous factors affect costs, estimating the cost of any particular mine must be considered within its own special situation and requirements. Some of the major variables are: type of mine, seam depths and thicknesses, types of coal to be produced, the rate at which the coal is to be mined and the desired rate of return on investment.

Katell (1976a and 1976b, pp. 1-37 and 1-41) estimates basic capital investment for both surface and underground mines assuming different production rates. In estimating capital investments for underground mines, costs for materials and equipment were based on 1975 cost indices. A fairly complete list of basic capital investments is presented in the references. A summary of the capital investment requirements for a 72-inch thick bituminous coal bed (seam) is presented in Table 5.3. The data in Table 5.3 further indicates total capital investment increases as the annual production rate increases. It is significant, however, that as the production rate increases, the capital investment per ton of coal produced decreases indicating economies of scale (See Manpower, Chapter 7).

Basic capital investment estimates for surface mines by region, costs for material and equipment were based on 1975 cost indices. A list of basic capital investments for strip mines is summarized in Table 5.4. Again, the data indicates capital investment per ton of coal produced decreases as the production rate increases. However, it can also be seen in the table that total capital investment for strip mines is more dependent on location than on production rate, while the investment for underground mines is more dependent on production rate.

As an example how location of coal resources affects capital investment, coal in the Colorado region is different from that in the Northern Great Plains region (which includes Wyoming and Montana). Generally, the Colorado coal regions have thinner coal beds (3-20 feet compared to a typical thickness of 25 feet in the Northern Great Plains region) and higher stripping ratios (the ratio of overburden thickness to coal seam thickness), which contribute to higher mining costs.

Tables 5.3 and 5.4 present operating cost data and the selling price of coal necessary to obtain a 15 percent, after tax, rate of return (discounted cash flow rate of return), again, as calculated by Katell. Discounted cash flow rate of return (DCFROR) is defined most commonly as the rate of return that makes the present worth of investment cash flow (including after-tax salvage value) equal to the present worth of all after-tax investments. This is usually the primary economic investment analysis used by industrial companies to evaluate the economic potential of investments (Stermole, 1974, p. 232). Similar to the capital investment costs per ton of coal produced in both underground and strip mines, the operating cost per ton also decreases as the production rate increases.

The use of a 15 percent DCFROR in calculating a selling price per ton is typical of today's investment decision analysis by coal companies. Although this rate may be a bit higher than some other industries, its selection reflects the judgment that the risks associated with the mining industry exceed those associated with rate-regulated industries such as transportation

Table 5.3

CAPITAL COSTS, OPERATING COSTS, AND SELLING PRICE BY ANNUAL OUTPUT CAPACITY FOR UNDERGROUND BITUMINOUS COAL MINES (1975 dollars)

	Million tons per year			
	1.06	2.04	3.18	4.99
Estimated initial capital investment	$20,799,800	$36,943,000	$54,853,800	$84,027,500
Estimated deferred capital investment	11,707,000	18,615,000	26,164,500	38,956,000
Total capital investment	$32,506,800	$55,558,000	$81,018,300	$122,983,500
Capital investment per ton of production	$ 30.67	$ 27.23	$ 25.48	$ 24.65
Operating cost per year	9,744,200	17,335,600	25,988,100	40,500,300
Operating cost per ton of production	9.19	8.50	8.18	8.12
Selling price per ton, 15 percent DCFROR	12.87	12.08	11.63	11.64

Source: Katell et al., 1976b, p. 5.

Table 5.4

CAPITAL COSTS, OPERATING COSTS AND SELLING PRICE BY ANNUAL OUTPUT CAPACITY AND LOCATION FOR SURFACE COAL MINES (1975 dollars)

	Million tons per year and location		
	4.8 Eastern	6.72 Interior	9.2 Northern Great Plains
Estimated initial capital investment	$65,137,900	$78,514,600	$42,030,100
Estimated deferred capital investment	21,832,000	30,863,900	36,493,400
Total capital investment	$86,969,900	$109,378,500	$78,543,500
Capital investment per ton of production	$ 18.12	$ 16.28	$ 8.54
Operating cost per year	20,992,000	25,696,400	24,027,300
Operating cost per ton of production	4.37	3.82	2.61
Selling price per ton, 15 percent DCFROR	6.94	6.03	3.39

Source: Katell et al., 1976a, p. 5.

(Federal Energy Administration, 1976c, pp. 4-5).

Table 5.5 compares capital costs, operating costs, total production costs, cost per ton, and manpower needs between underground and strip mines with the same production rate. From the data shown, it is clear that underground mines require more capital investment and have higher operating costs than strip mines. This, of course, contributes to higher prices for coal mined underground. Generally, underground coal sells for about twice the cost of surface-mined coal. Thus, underground-mined coal has to be of sufficiently high quality (e.g., higher Btu content, lower sulfur content, or lower ash or moisture content) to make it desirable at this higher price.

FINANCIAL FACTORS

Market Considerations

After a potential mining operator has assembled and carried out a minable coal package (a program for coal mine development including lease acquisition; drilling program development--drilling, sampling, logging, and analysis; and determination of commercial coal quantities present), he must complete the market development process of finding a customer or customers to purchase the coal. The typical market development process is outlined by Jones (1977).

The first step in developing a market is to conduct a market survey, which begins as work on the minable package is nearing completion. In addition, information about prospective new power plant locations, potential transportation methods, and environmental regulations affecting the use of the coal in these potential market areas is important to market analysis. Data obtained from a market survey is then analyzed to identify the best prospects. At the same time, potential customers will probably have completed a similar survey of potential suppliers looking for the type and amount of coal that best meets their needs.

Before a power plant boiler system can be designed, the characteristics of the coal to be burned must be known. Potential buyers and sellers probably exchange additional information about mutual needs.

Currently, the electric power utility industry has few commitments for new coal plants or plants which will convert to coal because of the uncertainty of future demands for electricity, air pollution restrictions, and regulations influencing the expense of converting oil and gas fueled power plants to coal. All these factors contribute to the slower development of new coal mines.

If arrangements appear satisfactory to both buyer and seller, they usually draw up a Letter of Intent to sell and purchase coal, begin contract negotiations, and continue until the contract is finalized. These contracts usually include provisions for modifications in the future (perhaps due to new

Table 5.5

SURFACE AND UNDERGROUND MINE COST COMPARISON (1975 dollars)

	Surface[a]		Underground[b]	
	Total	Cost/ton	Total	Cost/ton
Production, tons per year	1,000,000		1,060,000	
Capital Investment	$10,820,397	$10.82	$23,629,500	$22.29
Employees	49		246	
Annual Production Costs:				
Direct cost:				
Labor	$ 726,944		$ 2,928,900	
Supplies	987,000		1,955,400	
Power & water	300,000		406,100	
Union welfare	770,000		1,105,900	
Royalty	175,000		---	
Payroll overhead	290,778		1,171,600	
Total direct cost	$ 3,249,722	$ 3.52	$ 7,567,900	$ 7.14
Indirect cost:	$ 257,092	$ .26	$ 732,600	$ .69
Fixed cost:				
Taxes & insurance	$ 197,770		$ 407,000	
Depreciation	626,775		1,231,000	
Deferred expenses	269,785		---	
Total fixed costs	$ 1,094,330	$ 1.09	$ 1,638,800	$ 1.55
TOTAL ANNUAL PRODUCTION COSTS	$ 4,601,144	$ 4.60	$ 9,938,500	$ 9.38

Sources:

[a] U.S. Bureau of Mines, 1972, pp. 69-77. (Costs were updated using 1975 wage rates, mining equipment cost indices, and additional information from Bureau of Mines personnel.)

[b] Katell and Hemingway, 1974, pp. 6-14.

laws or regulations). The entire negotiation process is complex and may last a year or more.

The Federal Energy Regulatory Commission, formerly the Federal Power Commission, has compiled data on delivered coal prices. Sample prices are published monthly by the Bureau of Labor Statistics, and f.o.b. ("free on board"--price of coal at the mine before transportation) prices per ton are reported to the Bureau of Mines by coal producers (U.S. Bureau of Mines, 1976, p. 164). Only average prices for an area are available to the public; f.o.b. prices from individual coal suppliers are generally proprietary. Of course, the value and price of coal obviously differ considerably depending upon the mining method used, mine location, quality and rank, as well as the amount and type of beneficiation used.

Taxes

Any plan for a potential coal mining operation must consider the effect of taxation long before development begins. Colorado coal operators are subject to many taxes, including: (1) ad valorem property taxes, (2) State income taxes, (3) sales and use taxes, (4) royalty payments on State and federal lands, (5) excise taxes, (6) license and permit fees, (7) federal income taxes, (8) corporation organization and qualification fees, (9) corporate franchise taxes, and (10) unemployment insurance contributions (Spence, 1976, p. 95 and Natural Resources Journal, 1976, p. 435).

In addition to these tax considerations, the controversial Colorado Severance Tax took effect January 1, 1978, imposing taxes on metals, molybdenum, and fuels such as coal, and oil shale. It increased the previous tax on oil and gas, but it does not tax nonmetals such as sand and gravel.

Coal is taxed at a rate of 60 cents per ton for surface-mined coal and 30 cents per ton for underground-mined coal, with a provision for increasing or decreasing the rate 1 percent for every 3 points change in the index of wholesale prices. The first 8,000 tons produced each quarter are exempt and there is an additional 50 percent credit on the production of lignitic coal (Colorado Mining Association, 1977, p. 184).

Some 25 to 30 states have enacted severance tax legislation affecting one or more types of mineral extraction; the tax theory behind such legislation remains controversial. There are numerous arguments protesting and supporting the severance tax (see, for example, Spence, 1976, p. 94), but it is evident that the main purpose of a severance tax in Colorado (as is perhaps the purpose of all taxes) is to generate revenue. Local governments receive a portion of this revenue to help offset the adverse impacts caused by mineral evelopment.

It appears likely that most or all of the severance tax payment costs made by the coal mining industry will be passed on to the consumer. Unfortunately, marginal coal operations may be forced to halt production because of this additional tax. Murray (1978a, p. 2) estimates that the new Colorado severance tax will generate more than five million dollars from 1978 coal production. The continued influences of this additional revenue on the Colorado economy is expected to be significant.

Labor Unions and Productivity

Labor unions, in general, have exhibited significant power in the labor market in recent years. This has probably resulted in higher wage increases than would have developed in the prevailing labor supply-demand relationship. The United Mine Workers (UMW), a union representing approximately 40-50 percent of the mine workers in Colorado, has been active in exerting labor market power. Besides the UMW, the Redstone Union represents miners in several mines in northwest Colorado. The unions have been very effective in establishing prevailing wages and benefits for all coal miners. Even those mines employing non-union personnel have wage and benefit structures similar to those of the UMW, so union wage or benefit increases can be expected to affect both union and non-union coal mining operations.

Escalator clauses, normally built into union contracts, allow workers to keep up with inflationary cost-of-living increases. At the same time, a strong case can be made that escalator clauses may be one of the major factors fueling inflationary expectations. Regardless of point of view, mine operators expect continued labor cost increases and build such increases into their financial structure. A potential or existing mine operator must therefore consider the labor union and its influence on wages as an important aspect of commercial development.

In 1969, the amount of coal produced per man-hour in the United States was 2.03 tons for underground mining and 4.61 tons for all types of surface mining, resulting in a figure of 2.59 tons for all mining methods and 2.40 tons overall (if man-hours used for mechanical cleaning and independent shop and yard work are included).

Since 1969, however, productivity for coal mining has decreased. In 1977, productivity was 1.12 tons for underground, 3.43 tons for all types of surface, 1.91 tons for all mining methods, and 1.72 tons for overall labor (Allen, 1978). On the bright side, the national productivity figures for 1976 and 1977 show a slight increase over the values for 1975. Colorado productivity of 2.13 tons in 1977 for all mining methods was slightly higher than the national average.

A number of reasons contributed to the decline in productivity after 1969. These include enactment of the Federal Coal Mine Health and Safety Act (CMHSA) of 1969, rising costs of materials and labor, environmental legislation, deeper and more difficult to mine coal reserves, short supply of adequately trained mining personnel, and union-imposed work rules. These factors have influenced both underground and surface operations, but, most notably, the CMHSA has changed underground mining. The act has most certainly made coal mining a safer occupation through stringent rules concerning dust exposure, more secure roof supports, monitored and grounded power distribution systems, increases in underground ventilation, and more (Cornett, 1976, p. 120) (See Mining Safety, Chapter 7). However, these impacts will affect new planned mining activities less than existing activities, which were often designed to meet different health and safety standards (Federal Energy Administration, 1976b, p. 30).

Generally, strip mining operations have avoided declines in productivity because they are highly mechanized, they require fewer men than underground mining, and they have increased the rate of coal production through the use of the stripping shovel and dragline (Risser and Malhotra, 1976, p. 480). Since coal mining productivity depends on so many variables of health and safety, environmental production and regulation, opinions throughout industry and government differ over the productivity rates for the future.

Reclamation Costs

Land reclamation costs vary for different regions of the country and for specific sites within a region. Therefore, only a general estimate of costs can be made without considering such specific factors as type of terrain, soil, vegetation, type and thickness of overburden, coalbed thickness, ground and surface water, climate, size and type of equipment used, method of mining, reclamation laws and regulations, and the individual operator's method of reclaiming the land.

Persse (1977, pp. 1-34), in a recent survey of reclamation costs in the western United States, has developed some general cost estimates. In doing so, he divided the states west of the Mississippi into four regions, designated A, B, C, D; Colorado's coal areas are within Region C. Reclamation costs for each region were presented in four categories: (1) design, engineering, and overhead; (2) bond and permit fees; (3) backfilling and grading; and (4) revegetation. In his analysis, however, Persse did not include the costs of monitoring environmental variables and conducting site surveys. Thus, total reclamation costs are now greater than those presented in his analysis.

In general, total reclamation costs in Region C ranged from $1,670 to $5,040 per acre. The Routt County area of Colorado had reclamation costs ranging from $1,670 to $2,810 per acre, with 25 to 38 percent of these costs going for design, engineering, and overhead; 1-2 percent for bond and permit fees; 60-70 percent for backfilling and grading; and only 1-2 percent for revegetation. Again, since more current revegetation regulations have been enacted since Persse's

Plate 5.1 A scraper "top-soiling" in preparation for land reclamation at the Edna Mine in Colorado. (Courtesy: J.A. Pulver, Gulf Oil Corporation)

survey took place, the percentage of revegetation expenses has increased substantially for the western states.

Reclamation costs per ton of coal mined for the Routt County area range from 16-18 cents. Table 5.6 shows sample calculations converting reclamation dollars per acre to dollars per ton. A sample plot of coal seam thickness versus reclamation costs per ton is displayed in Figure 5.1.

The more stringent requirements for land reclamation brought about by the Surface Mining Control and Reclamation Act will cause some financial impacts on mine operations. Peabody Coal Company estimates the impact of implementing the new regulations to be as low as $1 per ton in the West (which often has existing programs that parallel the federal performance standards) and up to $8 per ton in states where small mines have had less regulation (Mining Engineering, 1978, p. 516) (See Labor Costs, Chapter 5).

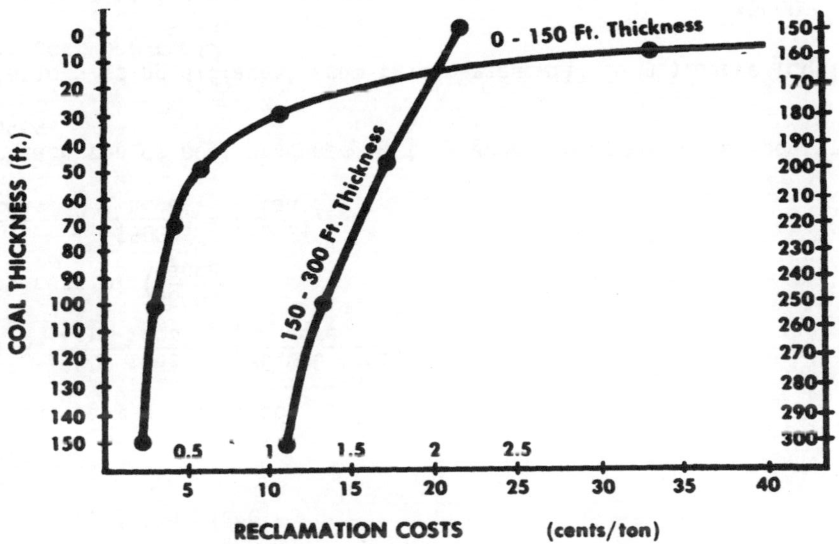

Figure 5.1 Strip mining reclamation cost - $6000/acre.
Source: Stanwood, 1978, p.7.

Table 5.6

CONVERSION OF RECLAMATION COSTS (\$/acre) to (\$/ton)

1) Density factor for coal "in place"

$$\left(\frac{1800 \text{ tons}}{\text{acre-foot}}\right)$$

2) For a 20 foot coal seam being mined,

$$(20 \text{ feet}) \times \left(\frac{1800 \text{ tons}}{\text{acre-foot}}\right) = \left(\frac{36{,}000 \text{ tons}}{\text{acre}}\right)$$

3) For reclamation cost of $\left(\frac{\$6000}{\text{acre}}\right)$,

$$\left(\frac{\text{acre}}{36{,}000 \text{ tons}}\right) \times \left(\frac{\$6000}{\text{acre}}\right) = \left(\frac{16.7¢}{\text{ton}}\right)$$

4) Therefore, for each ton of coal produced in this area, 16.7¢ would be spent for reclamation costs.

5) Similar calculations using different seam thicknesses will form a curve (coal thickness vs. reclamation cost per ton).

6) Other curves can be made for other reclamation costs per acre (e.g., $\frac{\$5000}{\text{acre}}$, $\frac{\$4000}{\text{acre}}$, etc.)

Source: Stanwood, 1978b, p. 9.

6
Environmental Considerations

Interest and concern for the environmental aspects of energy development have become more important nationally and locally only in recent years. In the case of coal, environmental concerns begin with the exploration phase of searching for a suitable mining site and extend to the removal of noxious waste products from burning coal. Between the alpha and omega of environmental concerns are mining (drainage, subsidence possibilities and surface reclamation), reclamation of water used for cleaning (beneficiation), transportation (pipelines), cooling power plants and coal conversion technologies, consumptive use of water (gasification and liquefaction of coal) and air quality (development, cleaning, load-out, transportation, and burning processes).

ENVIRONMENTAL LEGISLATION

It became increasingly clear to many of our national leaders by the late 1960s that some form of environmental legislation was needed to promote and preserve environmental quality. However, the debate concerning the ground swell of public concern for environmental preservation that came out of the 1960's, and continues today, has not settled the question about what is an optimum level of environmental protection. The four major pieces of federal environmental legislation which have had a major impact on all aspects of coal mining, processing, and utilization include: the National Environmental Policy Act (NEPA) of 1969; the Clean Air Act (CAA) of 1963, amended and expanded in 1967, 1970, and 1977; the Federal Water Pollution Control Act (FWPCA) of 1948, amended and expanded in 1965 and 1972; and the Surface Mining Control and Reclamation Act (SMCRA) of 1977.

Perhaps NEPA's major impact has been to require all federal agencies to prepare an Environmental Impact Statement (EIS) detailing the environmental impacts of proposed actions (initiated with leasing federal lands for coal development).

The CAA amendments of 1970 and 1977 have established strict limits on air pollution emissions; coal-burning plants emitting sulfur oxides must limit those emissions to 1.2 pounds of SO_2 per million Btus of coal burned. The FWPCA amendments of 1972 have established strict limits for potential water pollution, which although not as much of a problem for coal processes, still are very significant for the industry.

The SMCRA was enacted to protect against potentially adverse effects of surface coal mining and to fund the rehabilitation of lands left scarred by abandoned mining operations. The Abandoned Mine Reclamation Fund, created for mined-land reclamation, is primarily derived from fees (of 35 cents per ton of surface-mined coal, 15 cents per ton of lignite) paid by mine operators. User fees on reclaimed lands, donations, and funds from the sale of reclaimed lands are also deposited in the reclamation fund (Lusk, 1978, p. 139). A Federal Office of Surface Mining (OSM) has been established to enforce and regulate the provisions of the SMCRA.

Colorado is one of a number of states that enacted State laws to control air and water pollution and to provide for mining reclamation. The State air and water pollution regulation grew out of the Federal CAA and FWPCA, which recognize pollution control authority in individual states once the State plan is approved by the Environmental Protection Agency (EPA).

The Air Pollution Control Division and the Water Quality Control Division of the Colorado Department of Health are the responsible agencies for Colorado air pollution and water pollution control. The Colorado Mined Land Reclamation Act of 1976 established the State Mined Land Reclamation Board to review mining and reclamation proposals and problems and to develop standards for reclamation efforts. The Board is now a part of the Colorado Department of Natural Resources. The Colorado reclamation law is not as strict as the more recent federal law and in order to retain State control over reclamation practices, Colorado must enact legislation equal to, or more strict than, the federal standards by August 1979.

Many additional provisions of the environmental laws mentioned here have not been discussed. More specific information can be acquired from a study of the laws themselves or from the appropriate federal or State agencies.

ENVIRONMENTAL PLANNING

Planning to reduce the environmental impacts resulting from coal mine development has become an accepted part of the overall mining plan. There is no question that this involves considerable effort, time, and expense; however, careful and thorough environmental planning completed in the early stages of development will usually save additional effort, time and expense later, and will help to keep the environment as natural as possible. Jones (1977, p. 135) has explained the typical environmental study process for a new mine.

The first step consists of examining the vegetation, soils, topography, and drainage patterns and obtaining a photographic record of the area. The operator must examine existing environmental data, particularly if he is addressing any potentially serious environmental problem which may prohibit mining in the area altogether. Next, the operator must define the scope of work to be performed along with a time schedule and projected costs. It is necessary to consult with various governmental agencies at this stage to make sure all areas of environmental concern are properly considered.

Implementation of impact studies can now begin. This usually includes, as a minimum: environmental baseline studies; socio-economic studies; archaeological studies; and environmental analysis and mitigation. The end product here will be an Environmental Impact Report (EIR), which is submitted to appropriate regulatory agencies who issue permits or approvals. An Environmental Impact Statement (EIS) may also be required if the National Environmental Policy Act (NEPA) applies to the situation.

The EIS is first prepared in a draft version, and is reviewed and discussed in public hearings. Then it is prepared in a final version and submitted to the Council on Environmental Quality (CEQ) before approval for a mining and reclamation plan is issued. Normally the environmental quality is monitored throughout the life of the mine so that adverse impacts can be identified and corrected before they become irreparable.

Other coal facilities (such as power plants, beneficiation plants, coal conversion plants, etc.) will follow similar planning procedures to keep environmental impacts to a minimum, and to meet government regulations. Necessary pollution control equipment is usually best integrated into the engineering scheme and planning phase before plant construction begins so that costly delays do not occur later.

ENVIRONMENTAL PROBLEMS

Mining and Reclamation

No matter which mining methods are used to extract coal, significant changes may occur in the surrounding environment. The resulting environmental problems, however, differ greatly depending upon the mining methods that are used.

The most obvious effect of strip-mining coal is the scarred land which must be rehabilitated in order to return it to a productive and/or pre-mining conditions of the land. Beyond being aesthetically unpleasing, surface-mined areas cause temporary problems to plant and animal life within the surrounding ecosystem; and great care must be taken to insure that the entire ecosystem is not destroyed.

Reclamation is the process of backfilling, grading, and revegetating mined-out areas. It has become almost a landscaping art in some parts of the world where reclamation

Plate 6.1 Restoration of mined lands at Energy Fuels Corporation's surface coal mines is shown in this aerial photo of Pit No. 3. The smoothed area at right has been filled and restored to its original contour after removal of the coal. (Courtesy: Energy Fuels Corporation)

practices have been established for many years. While it is a relatively new process in the United States, it has already achieved tremendous success in the eastern and southern states where water is plentiful. Reclamation in the western states, however must overcome many problems to be successful: lack of water, high altitude soils, severe climatic changes, and incomplete knowledge of the affected ecosystem.

About 75 percent of all western coal fields receive less than 20 inches of precipitation annually. In addition, seasonal temperatures may vary from 50 degrees below zero to 120 degrees above zero F. There are usually only short frost-free periods. Topsoil is often inadequate, since western coal lies beneath a layer of geologically young material subject to excessive erosion. To further the problem, flash flooding and wind erosion not only increase the loss of valuable soil, but cause air pollution problems in the form of dust over vast areas (Gage, 1977, p. 20). Many research efforts are currently underway to help solve these problems.

Other environmental problems caused by strip mining include noise pollution from the mining operations; land erosion (which increases silt build-up in streams and lakes, possibly destroying wildlife habitat); the formation of sulfuric acid (from reaction of acidified water with sulfur in the coal) which may leach into streams and rivers; suspended solid water pollutants (resulting from solid waste pile runoff); air pollution (from diesel equipment and dust); and the problems of safely disposing of solid wastes (including sludge and overburden). These issues can be reduced significantly with careful planning and proper control procedures throughout the life of the project.

Environmental problems from contour mining or augering appear to be similar to those of area mining. However, in four environmental impact categories (water pollution, air pollution, solid wastes disposal, and land requirements), area mining usually produces fewer residuals than either contour mining or augering (University of Oklahoma, 1975, pp. 1-53).

Underground mining appears to have less environmental impacts than strip mining, but problems can still result if environmental factors are not carefully considered. The major problem here may be land subsidence caused by collapse of mined-out areas. This may occur during or shortly after mining, but usually happens many years after production has ceased. The longwall mining method, which allows the roof to collapse evenly as mining progresses, generally results in less harmful subsidence than the room and pillar method. Even slight surface subsidence can cause extensive damage to nearby buildings or structures. The Department of the Interior estimates that coal extraction has undermined eight million acres; to date, two million of these acres have subsided, creating land use problems (Federal Energy Administration, 1976a, p. 8).

Other environmental problems of underground mining include acid mine drainage and other water-soluble pollutants, which may contaminate streams and rivers, if not properly controlled; coal dust within the mine is hazardous to miners' health; disposal of solid wastes is also a problem. And, since Colorado has extensive underground coal resources, environmental problems related to development will continue to be active issues.

Beneficiation

The washing process is the main source of environmental problems from coal beneficiation techniques. Refuse pile runoff (after washing) is a major source of air and water pollutants, including acids, and, possibly, toxic trace elements. If left uncontrolled wind and water erosion usually speed the runoff process contaminating water sources and causing a local pollution problem. Settling ponds are used to control refuse pile runoff.

Disposing of solids generated during washing presents another major obstacle to protecting the environment. A typical washing plant, processing 500 tons of coal per hour, must dispose of about 1,000 tons of refuse daily, which is equivalent to about as much refuse as is generated by a city of 400,000 people each day (University of Oklahoma, 1975, p. 1-67).

Since Colorado coals, which are low-sulfur coals, require little or no washing at the present, these environmental problems are not as serious to the State as they are in other parts of the country. In an effort to remove undesirable substances in coal such as ash and sulfur, however, coal beneficiation processes are expected to have an increasing role in the coming years. In fact, coal washing is now the least expensive way of removing a given portion of the pyritic sulfur from coal (Gage, 1977, pp. 22-23). With this in mind, some observers believe the potential environmental problems from beneficiation may soon have a large impact in Colorado.

Burning

The primary chemical emissions from burning coal include five air pollutants: carbon dioxide (CO_2), carbon monoxide (CO), nitrous oxides (NO_x), sulfur dioxide (SO_2) and fine particulates, i.e., solid wastes suspended in the air. These air pollutants are, to the extent they are present, hazardous to human health as well as to agricultural crops. Significantly different amounts of these pollutants result from the burning of different grades of coal, especially in relation to sulfur content in coal. Use of a low-sulfur coal may save a coal-burning facility considerable expense for SO_2 controls.

Sulfur

All utilities and industrial users of coal are required to comply with stringent state and federal regulations concerning air pollutants. This has been a matter of intense debate since all coal burning facilities across the nation have been included under these regulations.

Many utility companies claim that the costs of such regulations outweigh their benefits; therefore restrictions should be eased so the United States can utilize the energy in the nation's vast coal resources. Others argue that we must enforce strict regulations to protect against an air pollution catastrophe, which may become increasingly possible with more widespread burning of coal. The ultimate solution to this debate is uncertain, but it appears that environmental regulations are here to stay.

Elimination of SO_2, the noxious combustion product in sulfur-containing coals, may be accomplished: (1) prior to combustion--coal cleaning for iron pyrite removal and/or the use of leaching or of solvents--solvent refined coal; (2) during combustion--fluidized bed processes; (3) flue gas desulfurization (FGD)--stack gas cleaning; or (4) some combination of these methods. Performance and costs (including energy costs) are the major deciding issues.

Fluidized bed combustion presents an opportunity for removing sulfur from coal during combustion, enabling the direct burning of high sulfur coals. The process requires a bed containing ash, powdered dolomite or limestone supported on a distributor plate. The bed is fluidized during combustion by forcing air through the plate. Sulfur combines chemically with the limestone, converting it from a carbonate to a sulfate (which is stable at combustion temperatures). This sulfonated sludge is then removed with the ash. Federal Government and industry are doing considerable research and development on this process, and prototype testing is underway to determine performance parameters and costs.

Air Pollution Emission Controls

Gage (1977, pp. 21-25) explains the specific pollution control technologies for the three major air pollutants as follows (the emphasis here is on controlling air pollution emissions).

There are many air pollution control devices in use today in a coal burning facility. One of the most common of these devices is the Flue Gas Desulfurizer (FGD), often called a scrubber. The function of scrubbers is to remove pollutants (mainly SO_2, but also NO_x and to a lesser extent particulates) from flue gas emissions. This is usually done by means of a liquid or solid sorbent. Today there are more than a dozen reputable firms selling various types of scrubbers; most of these units can remove at least 80 percent of the sulfur oxides from power plant emissions.

Generally, there are two basic types of "wet" scrubbers available. One produces a sludge which must be disposed, while the other produces a by-product which can be sold. Lime or limestone scrubbers are by far the most common among the sludge-producing types, accounting for approximately 75 percent of scrubbers installed or scheduled for installation at present. Another major sludge-producing system is the dual alkali scrubber; it requires less than half the energy required of other flue gas desulfurization processes now being installed.

Sludge removal, with costs not yet well defined, is a serious environmental problem to be considered in an effective pollution control program. Scrubbers which produce sulfuric acid or elemental sulfur as a by-product are being given serious consideration for development.

Developing technology appears to show that "dry" scrubbers which use solid nahcolite (a saline mineral, naturally occurring sodium bicarbonate, and found extensively with oil shale, particularly in Colorado's Piceance Basin where resources of thirty billion tons have been identified) may be more effective than wet scrubbers. Nahcolite can remove about 90 percent or more of SO_2, and about 40 percent of NO_x, all with lower replacement and operating energy costs (Weichmann, 1976). With this market in mind, some oil shale developers are planning to recover nahcolite and other saline minerals in conjunction with shale oil.

By mid-1977, Gage reports that 51 electric power companies had installed, were installing, or were planning to install some 122 scrubber systems, nationally. When completed, these 122 scrubbers will control the SO_2 emissions for a power-generating capacity of nearly 50,000 megawatts. The nation will be more than halfway to its 1980 goal of 90,000 megawatts of scrubber-controlled electric generation (which the EPA estimates will be needed to meet primary SO_2 emission standards by late 1980). The capital and operating cost impacts resulting from scrubber use are listed in Table 6.1.

Two other major air pollutants, particulates and nitrous oxides, also require specific control procedures and equipment. For particulate control, the modern high-efficiency electrostatic precipitators (ESPs) can effectively control particulate emissions. However, their effectiveness drops if the ash content from coal has high electrical resistance, which is typical of ash from most western coals. While it appears that use of western coals may help to reduce the SO_2 emission problem, it creates an increased particulate emission problem. Research today is centering on use of scrubbers, fabric filters (so-called bag houses) and changes in ESP design to control particulate emissions in the combustion of all coals.

Nitrous oxides (NO_x) present still another challenge for emissions control. The only method consistently successful today for NO_x control is combustion modification, a process which uses changes in the combustion conditions within the boiler to minimize high-temperature fixation of atmospheric

Table 6.1

EMISSIONS CONTROL IMPACT - OPERATING COST AND HEAT RATE PENALTY (500 MW Coal-Fired Plant)
(1976 dollars)

	$/KW	Equipment Cost ($)	Op. + Maint. Cost Mills/KWh	Heat Rate (Penalty)**	Control Capability
Base Power Plant*	700	350	12*	9,000	-----
Particulate Control	50	24	1	(90)	0.03 lb/MMBtu 99% removal efficiency
SO_2 Control	150	75	7	(900)	90% removal efficiency
NO_x Control	60	30	4	(90)	0.2 lb/MMBtu
Totals	$ 960	$ 480	24	10,080***	
Penalty	(+37%)	(+37%)	(+100%)	(-12%)	

Source: Yeager, 1978, p. 135.

* Includes fuel cost
** ( ) denotes penalty
*** Heat rate required to achieve base power

nitrogen with oxygen. In current coal-fired power plants, combustion modification reduces NO_x emissions by only 40-50 percent, and there are serious questions about the effect of these modifications on the long term operation and life of the boiler. Current research in controlling NO_x centers on using scrubbers and further modifying the combustion process.

Conversion

Future uses of coal as a major energy source in the world may be primarily through conversion processes such as gasification and liquefaction (See Chapter on Technological Factors Affecting Coal Use). Although commercial development of these processes has only reached the pilot plant stage in the United States, operational systems in future years appear inevitable.

The major pollutants from an above-ground gasification facility would include SO_x, solid wastes, as well as various water effluents which require control through water recycling and evaporation ponds. Gasification processes, because of more efficient pollution control technology, may have a significant environmental advantage over direct burning of coal. Emissions from the coal gasification and combustion are compared in Table 6.2. This fact seems especially relevant in the discussion of the troublesome SO_x emissions from coal burning plants. Established technology in the gasification process removes more than 99 percent of SO_2 from stack gases, hence, there should be no significant SO_2 problem with gasification.

A liquefaction facility would be faced with pollution problems similar to those of gasification. But again, it appears that the liquefaction process may also have an environmental advantage over direct burning of coal. This conclusion must be qualified, however, since not enough data about liquefaction pollutants have been compiled.

Solvent Refined Coal

Other potential options for pollution control are processes which pretreat coal prior to combustion and which produce solid and liquid coal products. These processes, producing what is called solvent refined coal (SRC-I and II), may be used for high-sulfur bituminous coals; technologies are sufficiently advanced to enable construction of demonstration plants (Jackson and Schmid, 1978, p. 331).

In the first process (SRC-I), coal is dissolved in a recyclable solvent, in the presence of hydrogen and at an elevated temperature and pressure. The undissolved portion of the coal, primarily ash, is removed by filtration. Vacuum distillation recovers the recyclable solvent leaving solid, low-ash coal. Technical difficulties which could be resolved only at high cost could result when SRC-I is converted to a commercial size operation. These concerns led to the

Table 6.2

SUMMARY COMPARISON OF ENVIRONMENTAL IMPACTS OF TWO ENERGY-EQUIVALENT PROJECTS

	High-Btu Coal Gasification Plant (250 mmcfd)*	Kaiparowits Power Plant (3000 Mwe with scrubbers)
Air Emissions (LB/HR)		
Particulates	180	1,070
SO_2	450	4,300
NO_x	1,780	20,830
CO	90	1,200
HC	30	360
Solid Wastes (Tons/day)	1,400	5,100

Source: American Gas Association, 1977, p. 7.

*MMcfd = million cubic feet per day

development of SRC-II which uses more extensive hydrogenation and hydrocracking processes. The SRC-II technique yields a distillate fuel oil with sulfur content reduced from about 3.0 to 0.25 by percent weight, and with heating values increased from 11,500 to 17,300 Btus per pound (Ibid., pp. 337-340).

The advantages of using solvent-refined coal, particularly SRC-II, are the minimal burner modifications and lower capital and operating costs (especially those related to the use of scrubbers, fuel transportation costs, and ash disposal) for power plants. Further, these plants could be on-line several years ahead of those using liquefaction processes (Harrison, 1978, p. 1007).

Water Availability and Use

Perhaps the one factor which may limit energy development in the West is lack of available water. Since all energy technologies require water resources, the widespread expansion of coal mining, burning, and conversion in the arid lands of the West may be impossible simply because there is not enough available water (See Transportation, Chapter 7).

Examples of water use in energy technologies include cooling water for electric generating plants or coal conversion processes; and water use in coal mining and subsequent reclamation. In addition, since coal seams may be aquifers (underground storage and transportation zones for fresh water), mining of the seam may disrupt the aquifer and change the distribution of underground water supplies in the area.

Most of the water used by energy technologies is fresh water. And, considerable uncertainty exists as to the long-term, physical and legal availability of water in Colorado and the other western states. This is especially the case in the Colorado River Basin drainage area (consisting of sections of Colorado, Utah, Wyoming, New Mexico, Nevada, California, and all of Arizona). Many interests, primarily agriculture, have current legal rights to these water supplies. Thus, water availability for many energy developments in this area may be impossible unless there are modifications to existing laws.

Many possibilities for water conservation in the energy technologies exist which could effectively lower water requirements. Table 6.3 shows water needs in various western locations for electric power generation, the Lurgi and Synthane gasification processes, the Synthoil liquefaction process, and a slurry pipeline. Although one would assume that industry would use the same criteria, published estimates of water use (especially for the synthetic fuel facilities) have tended to be much higher than those shown (Plotkin, 1977, pp. 229-230). However, water requirement figures vary considerably, depending upon who is doing the estimating and under what assumptions the estimates are made.

Other Environmental Impacts

A secondary environmental impact of coal-generated electricity (and all other forms of electricity generation as well) is the transmission and distribution system which requires overhead transmission lines, towers, and poles. Because these structures are not aesthetically pleasing to many people, careful planning as to their design and location is required, especially in undeveloped areas. Other negative visual impacts related to coal development include the mines, plants, and development structure. While these visual impacts can be reduced, it does not appear to be possible to remove all signs of a mining operation.

Environmental impacts from the surface and subsurface disturbances of coal exploration usually affect only a small area. If exploration is performed carefully, the impacts related to these activities can be kept limited and temporary.

Each coal transportation method has its negative impacts on the environment. These can include: water pollution from slurry pipelines; air pollution from unit trains and trucks; and land requirements for pipelines, roads and rail track. With effective planning and design, these negative impacts can be reduced, and perhaps eliminated. (See Colorado Impacts, Chapter 7, for more discussion of coal transportation.)

Table 6.3

WATER REQUIREMENTS FOR COAL CONVERSION TRANSPORTATION AND POWER GENERATION TECHNOLOGY BY SITE[a]

	Water Requirements[a] (1,000 acre-ft/yr)				
Site	Electric Power Generation	Lurgi	Synthane	Synthoil	Slurry Pipeline
Kaiparowits/Escalante	29.82	NC	NC	NC	NC
Navajo/Farmington	29.21	5.64	8.67	11.75	NC
Rifle	28.47	NC	NC	NC	NC
Gillette	25.84	4.21	7.78	9.23	19.17
Colstrip	26.66	4.62	7.81	10.30	NC
Beulah	23.88	3.31	7.67	10.09	NC

Acre-ft/yr = acre-feet per year
NC - not considered

Source: Plotkin, 1977, p. 230.

[a]For a 3,000 megawatt-electric power plant at 70-percent load factor, for 250 million cubic feet per day gasification facilities at 90-percent load factor, 100,000 barrels per day coal liquefaction facilities at 90-percent load factor and a 25 million tons per year slurry pipeline at 100-percent load factor.

Summary of Environmental Problems

The potential for some type of adverse environmental impact can be found in every aspect of coal extraction and utilization. Surface mining creates a need for land reclamation; underground mining may result in subsidence; coal washing may create water pollution; burning produces air pollution and solid waste disposal problems; coal conversion causes air pollution and water utilization; water requirements are large for each coal energy process; and the list goes on. Coal, indeed, lives up to its reputation as a "dirty" fuel; yet careful planning and management can significantly reduce its negative impacts.

If coal is to be used widely in the future, there will be many compromises with environmental issues. At present, no one knows whether extensive coal development is our best near-term energy alternative, or if another alternative is possible. However, the fact that coal is the most abundant energy resource in the United States cannot be overlooked in any analysis of the nation's energy needs. It is clear that coal will play a large role in our future energy needs, yet because of the many environmental impacts it creates, it will certainly never become a panacea.

7
Colorado Coal: Importance and Impact

PRODUCTION AND USE

Keith Murray (1978a, p. 2) reports that nearly 618 million tons of coal have been produced in Colorado during the 114-year period between 1864-1977. Great as this sum seems, it amounts to less than one year's output in the entire United States. Historically, nearly 70 percent of the State's coal production came from five fields in eastern Colorado; however, the trend gradually changed so that in 1976 less than 10 percent of the state's production was mined on the Eastern Slope. Figure 7.1 displays the historical pattern of coal production in Colorado.

In 1977, Colorado produced 11.97 million tons of coal, worth an estimated $200 million. Approximately two-thirds of this production came from surface mines. Of the one-third mined underground, over 80 percent was used for metallurgical purposes. Table 7.1 shows Colorado coal production by counties and number of producing mines for the period 1972-1977; Table 7.2 shows underground metallurgical coal production and Table 7.3, coal shipments out of State for the period 1960-1977.

Over two-thirds of the coal consumed in the State (including state-produced and imported coal) is used for steam-electric power generation. Table 7.4 shows Colorado's existing steam-electric power generating system, while Table 7.5 shows electricity generation in Colorado by fuel source. Coal is the most important electric power generating fuel source, accounting for 60 percent of the State's generating capacity. Most of the remaining coal consumed in Colorado is used for metallurgical and industrial purposes.

Murray further reports that 1978 coal production should be between 13 and 15 million tons, despite the coal strike earlier in the year. Thus, the record production of 12.6 million tons of coal, set in 1918, should easily be broken. Projections for 1985 range between 30 and 50 million tons per year, depending on development of the Denver basin lignites. Information on specific existing mines in Colorado can be

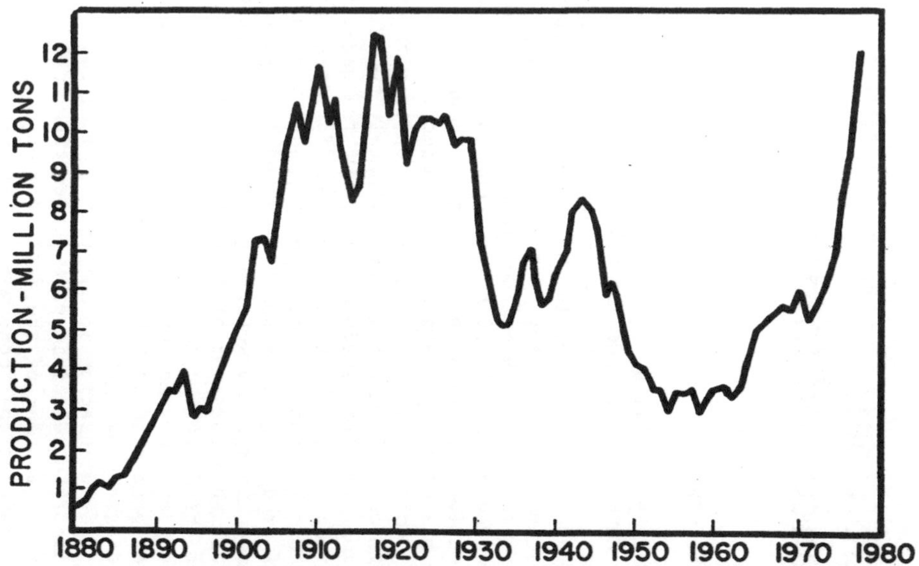

Figure 7.1 Historical Plot of Colorado Coal Production
Source: K. Murray, 1978.

TABLE 7.1

COLORADO COAL PRODUCTION (TONS) AND NUMBER OF MINES BY COUNTY (1972 - 1977)

County	1972	1973	1974	1975	1976	1977
Archuleta						4,070/1
Delta	31,433/3	18,256/1	5,583/2	642/2	14,023/3	327,352/5
Fremont	214,948/8	247,172/8	152,681/4	147,619/4	90,956/6	90,665/7
Garfield	2,231/1	2,106/2	1,807/2	1,835/2	1,425/3	70,793/7
Gunnison	786,081/4	1,038,113/4	1,259,536/4	1,298,778/4	1,246,723/7	1,347,182/6
Huerfano	5,243/1					
Jackson			7,899/1	320,677/3	270,085/3	495,956/2
La Plata	11,370/3	9,488/2	9,913/2	15,790/1	16,870/4	25,648/5
Las Animas	616,327/1	624,045/1	539,845/1	632,207/2	649,468/3	742,315/5
Mesa	11,275/1			75,738/1	57,134/3	300,199/3
Moffat	295,873/2	289,560/2	240,751/2	509,749/2	507,010/7	1,113,015/5
Montrose	92,698/1	106,798/1	105,723/1	104,980/1	97,929/1	94,402/1
Pitkin	652,352/3	780,602/5	865,988/5	926,546/5	889,520/5	936,430/7
Rio Blanco	4,556/1	329/1	11,766/1	2,871/1		8,836/1
Routt	2,231,985/5	2,606,025/5	3,457,899/5	4,073,351/7	5,553,486/9	6,309,173/10
Weld	574,707/3	509,952/2	300,295/2		66,874/1	105,103/2

SOURCE: Colorado Division of Mines Annual Statistics 1972-1977.

TABLE 7.2

UNDERGROUND METALLURGICAL COAL PRODUCTION
(thousands of tons)

Year	Las Animas	Pitkin	Gunnison	Total
1960	450	251	121	824
1961	761	383	129	1,274
1962	655	263	83	1,002
1963	749	387	49	1,187
1964	777	508	310	1,596
1965	854	604	372	1,830
1966	833	585	503	1,923
1967	745	629	687	2,062
1968	765	702	682	2,150
1969	763	667	722	2,154
1970	602	827	879	2,308
1971	747	509	917	2,174
1972	616	652	634	1,902
1973	624	780	886	2,290
1974	589	865	1,122	2,528
1975	632	926	1,163	2,722
1976	649	890	1,247	2,786
1977	742*	936	1,347	3,025

SOURCE: Colorado Division of Mines Annual Coal Statistics, 1960 - 1977.

1. 1975 production of underground metallurgical coal was 2.7 million tons, representing 33 percent of Colorado total coal production.
2. In-state consumption was primarily that used by OT & I in Pueblo.
3. Most of Colorado's metallurgical coal is exported to smelters in Montana, coking plants in Utah and eastern states.

TABLE 7.3

EXPORTS: COAL SHIPMENTS OUT OF COLORADO (thousands of tons)

Year	Exports
1960	750
1961	818
1962	578
1963	290
1964	1,094
1965	1,264
1966	1,330
1967	1,442
1968	1,525
1969	1,534
1970	1,537
1971	1,680
1972	1,303
1973	1,720
1974	2,858
1975	2,560
1976	2,760
1977	4,312

SOURCE: Colorado Division of Mines, Annual Coal Statistics, 1960-1977.

found in Appendix A. Further information can be obtained from the Colorado Geological Survey and Colorado Division of Mines. Information on proposed coal facilities (mines, power plants, conversion facilities, etc.) can be found in Rich (1978).

The amount and rate of coal production in a state depends upon lease availability, access to transportation, markets for the product, capital for equipment and mine development, and manpower and environmental concerns. In addition, forecasting Colorado's coal production is hampered considerably because there is no comprehensive correlation of basic geological data. Much of these data are now available, and their analyses by such organizations as the Colorado Geological Survey appear to be warranted. This information, when

TABLE 7.4

COLORADO EXISTING STEAM ELECTRIC GENERATING PLANTS

Company	Unit Name	Capacity MW	Primary Fuel	Alternate Fuel
CCS	Drake	282	G,C	O,G
CCS	George Birdsal	63	G	O
CUEA	Bullock	10	C	none
CUEA	Hayden	240	C	none
CUEA	Nucla	36	C	none
CUEA	Pagosa Springs	2	O	none
CTUC	Pueblo	10	G	unknown
CTUC	Pueblo New	31	G	O
CTUC	Rocky Ford	18	G	unknown
CTUC	W N Clark	44	C	G
CL	Lamar	37	G	O
PSC	Alamosa	49	O	G
PSC	Alamosa Terminal	29	O	G
PSC	Arapahoe	351	C	G
PSC	Cameo	75	C	G
PSC	Cherokee	802	C	G
PSC	Comanche	733	C	none
PSC	Fort Lupton	90	O	G
PSC	Fruita	29	O	G
PSC	Valmont	354	G,C,O	C,G
PSC	Zuni	115	O	G
TMPL	Trinidad	12	G	unknown
TSGTC	Burlington	117	O	unknown
TSGTC	Republican River	225	O	none

SOURCE: U.S. Department of Energy, 1977, pp. 247-250.

Key:

Companies:

CCS	City of Colorado Springs, CCS
CUEA	Colorado-Ute Electric Assn, Inc. CUEA
CTUC	Central Telephone and Utility Corp., CTUC
CL	City of Lamar
PSC	Public Service Company of Colorado
TMPL	Trinidad Municipal Power and Light
TSGTC	Tri State Generation and Transmission Assn.

Fuels:

C	Coal
O	Oil
G	Gas

TABLE 7.5

COLORADO POWER PLANT PRODUCTION BY FUEL SOURCE

Fuel	MW	%
Nuclear*	0	0
Hydro	691	16
Oil	721	17
Coal	2,630	60
Gas	305	7
Unknown	16	0
Other	0	0
Total	4,353	100

SOURCE: U.S. Department of Energy, 1977, p. 246.
*Ft. St. Vrain Reactor (330 megawatts)

developed, will provide a sounder basis for predicting the extent of coal production in the State. (Many of the constraints to coal production are treated in other chapters of this volume.)

TRANSPORTATION

Among the major considerations in analyzing how Colorado's coal will be used is an evaluation of the transportation alternatives for moving the coal—or the energy it may generate—out of the State. And in considering the influence of transportation on power generation costs, it is noteworthy that the Edison Electric Institute, an electric industry research organization, estimates that one-third of the average cost of coal-generated electricity can be attributed to transportation or transmission costs.

To meet the export demands for energy from Colorado steam coal, there are four major transportation alternatives:

- Transporting coal by rail in unit trains.
- Transporting coal by slurry pipelines.
- Generating electricity on site and transmitting the power.
- Converting to synthetic fuels (gas or liquid) and transporting by pipelines.

Key aspects of the technology involved in each transportation option are outlined as follows:

Unit Trains

Unit trains comprise a "dedicated" string of cars and locomotives operating on a regular schedule between a single origin and a single destination. A common unit train is 100 cars with each car having a 100-ton capacity. Loading and unloading can become accomplished in only a few hours since it usually is done while the train is moving. The economic attraction of the unit train is that, by remaining permanently assembled and dedicated to one route, it offers major savings over mixed-cargo trains.

Slurry Pipeline

The slurry pipeline involves transmitting coal which has been ground to a maximum of 1/8 inch in diameter and which has been mixed with 50 percent water (or other liquids) by weight (Office Technology Assessment (OTA), p. 31). In other words, when water is used as the transport medium, it takes about a pound of water to move a pound of coal.

The coal/water slurry is pumped through an underground pipeline similar to an oil pipeline. At the destination, the coal particles are dewatered, after which the coal may be burned. Since coal-fired boilers use ground coal, the small particle size is no problem. In principle, over one-half the water could be returned to the point of origin via an additional or a "dual flow" pipeline (OTA, 1978, p. 128). However, the process of returning the water to its point of origin is expensive. If not returned to the point of origin, the water could probably be used in the power plant's cooling system.

Slurry pipelines can use any quality of water, including water too saline for other beneficial uses. However, low-salinity water is preferred for the slurry, since some salts can be absorbed by coal, later causing corrosion, boiler scaling, and increased air pollution when the coal is burned.

Slurry pipelines are capital intensive and only economic (in comparison with unit trains) when large quantities of coal are moved from a central source area to a limited number of customers located close together, a requirement that strongly favors electric utilities. Consequently, ten to twenty-five million tons of coal per year are frequently suggested as economic slurry pipeline capacities (OTA, 1978).

Contrasts - Pipelines vs. Railroads

Fifteen slurry pipelines are operating worldwide, with two more under construction. All the slurry pipelines located outside the United States carry minerals other than coal. Nonetheless, the technology for coal slurry pipelines appears

Plate 7.1 Aerial photo shows Energy Fuels Corporation's Tipple No. 2, where unit coal trains are loaded by gravity as they pass underneath the stockpile. The crushing plant is shown at the left of the tipple. (Courtesy: Energy Fuels Corporation)

to be reasonably well-tested. For example, the Black Mesa coal pipeline in Arizona operates 99-1/2 percent of its available time. (Environmental Science & Technology, 1976, p. 1987).

At least five long distance coal pipelines are in the planning stage or under construction in the United States today. The Energy Transportation Systems, Inc. (ETSI) pipeline from Gillette, Wyoming to Baton Rouge, Louisiana by way of White Bluff, Arkansas, is expected to be completed first, with a service date of late 1979 if right-of-ways are secured. Chemical and Engineering News (1977a, p. 20) reports some interesting comparisons between the ETSI pipeline and a comparable rail transportation system:

The estimated cost of the pipeline is $750 million, assuming that the historical inflation trend continues. The comparable cost of a rail system to do the same job (carry 25 million tons of coal per year over the 1040 mile distance) would be about the same; however, this cost estimate does not include upgrading of the road bed for heavy-duty service nor the continuing maintenance and replacement of ballast and ties along the route.

The only construction material common to both the ETSI pipeline and the comparable railroad is steel. The pipeline will require 450,000 tons of steel while the railroad would use 850,000 tons. Energy costs for operating the pipeline are estimated at $5.6 million per year, compared to about $30 million per year for rail transportation.

Three hundred thirty-five people will be needed to operate the ETSI pipeline once it was constructed, whereas 2,570 would be needed on the railroad. Depending on the point of view, this could be a positive or negative point for pipelines. While the pipeline would save considerable expense for its developers, it would not employ numbers of persons that a railroad would. However, pipeline advocates counter this deficiency by noting that additional jobs will be created at both terminals of the pipeline.

Transportation of coal by pipelines has some advantages over railroads. Since the pipelines will be underground, they will not be visible except at the pumping stations. Unit train traffic through local communities causes considerable congestion, delays, and inconveniences, and has placed pressure on the railroads to reduce these impacts. In a study of Colorado State unit train traffic, it was concluded that by 1985, unit train traffic will increase to as many as 172 trains per week to and from coal markets along the Front Range. In total, there may be as many as 344 two-way trains in Colorado. The resulting congestion will probably require at least 25 new railroad and highway grade separations along the Front Range, at a potential cost of about $1 million each. Under Colorado's current Highway Action Plan, the State plans to build one grade crossing per year (Los Alamos Scientific Laboratory, 1978, p. 144).

Slurry pipelines, however, are not without problems or controversy. One problem is securing right-of-way across rail lines. The ETSI pipeline, for example, must cross existing rail lines at forty-nine places, and railroads are reluctant to waive their present rights-of-way (Chemical and Engineering News, 1977a, p. 20). ETSI appears to be making progress in obtaining rights-of-way, and President Carter is reportedly in favor of granting slurry pipeline companies the right of eminent domain so they may secure necessary right-of-ways (Gibson, 1978, p. 61). However, the slurry pipeline industry was handed at least a temporary setback when Congress voted down the slurry pipeline eminent domain bill in mid-1978.

Another problem for the pipeline is corrosion-erosion. This could occur in the pipeline itself from the gradual leaching of highly acidic compounds from coal which passes through the pipeline (Chemical and Engineering News, 1977a, p. 20). This may be controlled using chemical additives as neutralizers.

Perhaps the major problem for western pipelines will be the scarcity of water. Water unsuitable for municipal and agricultural use is being considered for pipeline use, but this may not be possible. Rail transportation, of course, requires only minimal amounts of water. Dual pipelines have been suggested so that the water used can be recycled, but the additional costs would eliminate many of the economic advantages of pipelines.

Transportation of coal by pipelines can become economically attractive with technological advancements. Recent studies of pneumatic pipelines for coal transport are proving favorable since the operating pressure does not have to be high and coal can range from dust size to pieces two inches in diameter. Even more important, air, rather than water, can be used to transport the coal, thus removing one of the main objections to slurry pipelines, although the possibility exists that static electricity could spontaneously ignite the coal dust if air is used (Los Alamos Scientific Laboratories, 1978, p. 150).

Other mediums for transporting coal in pipelines, such as the "methacoal" approach, hold promise. In this case, part of the mined coal is converted to a liquid such as methanol and then used for slurry transport. At delivery, the entire slurry can be burned directly as a high quality fuel, or the two can be separated at low cost (Gibson, 1978, p. 61).

It is clear that pipeline and railroad operators both have much to gain or lose in the struggle for coal transportation supremacy. But there really is no reason to think that the victory will fall to one or the other for both railroads and pipelines will be needed to meet the nation's demands for coal. Specific situations may favor one alternative over the other. Controversial decisions will be made by local, state, and federal legislators, but the decisions will be based on moving coal in an economical, efficient, and environmentally safe manner.

On-Site Electric Power Generation

Conventional electric generation involves burning coal (or other fuels) which generates steam to operate turbines. Condensing the steam for return to the boiler requires a large cooling system which usually involves water evaporation. As a side effect, emissions from the burning of coal cause air pollution, sulfur dioxide (SO_2) being an example of prime concern. If a scrubber is used to reduce SO_2 emissions, the resulting scrubber sludge will require solid waste disposal (See Environmental and Technology chapters for further data).

Synfuels

Coal-gasification and liquefaction are developing high capital-intensive technologies with relatively large water requirements. Water, in these conversion processes, is used as a process coolant and a chemical ingredient (it is a source of hydrogen to satisfy the chemistry of conversion). They are not expected to be in commercial operation for another ten to twenty years. Underground gasification, now yielding a low Btu gas, is attractive because it is cheaper, utilizes small, deeply buried coal deposits, appears to have potential for little environmental disturbances and higher Btu values than obtained in early research, and it could be available sooner than the preceding conversion methods (See Technology chapter for additional information). Low Btu synthetic gaseous fuels are not used for generating electric power now, however, technological advances may also make this use a potential alternative for the future.

Water Use

Consumption of water--as opposed to water withdrawals—is the key to estimating water requirements for coal development and transportation. This issue, probably the one of major concern to the state, is the only issue discussed at length here.

Unconsumed water may potentially be returned to the source, for additional beneficial uses, at added cost. Table 7.6 provides a comparison of water consumption for each transportation alternative. In the case of coal slurry pipelines, all the water withdrawn is shown as being consumed since it is only available for use at the destination. If a dual flow pipeline were installed, the water consumption could be reduced by at least 50 percent. However, it should be recognized that the dual-flow slurry pipeline is not without problems. These include additional capital investments and increased energy costs for pumping the water back to the point of origin.

Some discussion of potential slurry pipeline water use has suggested that water transported in pipelines might be

TABLE 7.6

WATER CONSUMPTION IN COAL UTILIZATION

Process	Water Consumption	Acre-Feet Per Trillion BTU Output (@ 74% utilization)
Unit Train	Small quantities for dust control	
Slurry Pipeline	Approximately 12,500 acre-feet/year for a 25 million ton per year pipeline of Colorado coal	24.3*
Electric Power Generation (1000 MW Plant)		
-Evaporative cooling	6,800-15,000 acre-feet/year/ 1000 MW unit	310-680
-Pond or river cooling	3,600- 8,150 acre-feet/year/ 1000 MW unit	165-370
-Wet/dry tower	500- 5,000 acre-feet/year/ 1000 MW unit	25-230
	100 acre-feet/year/ 1000 MW unit	4.5
Synthetic Liquid Fuel	5,800-130,000 acre-feet/year/for a 100,000 barrel/day plant (Lower portion of range reflects use of dry cooling)	30-680
Synthetic Gas	2,400-45,000 acre-feet/year for a million SCF per day plant (Lower portion of range reflects use of dry cooling	30-485

SOURCE: Taylor, 1976.

*For 10,000 BTU/lb coal.

considered a portion of water appropriation already due a downstream state. However, this issue is extremely complex and closely dependent on the quality of water that might flow out at the end of the pipeline. The issue requires considerably more study than this report can provide.

To place the water use in context, a 25 million ton/year coal slurry pipeline is expected to use about 12,500 acre-feet of water each year. As a comparison, this represents less than 0.20 percent of the virgin flow of the Colorado River system as it leaves the State of Colorado.

Most sources suggest that the water required to convert the same quantity of coal to electric power or synthetic fuels is several times greater than needed for slurry pipelines, but this would depend largely on plant design (OTA, 1978, p. 110). A little water is used for dust suppression during the loading of unit trains. Water use by urban development, associated with synthetic fuel plants and power stations, is usually a small percentage of that used by these plants.

As examples of the water consumption requirements for various energy development or transport systems, it is estimated that (at a 74 percent utilization factor) a slurry pipeline would consume 24.3 acre-feet per trillion Btus of energy moved; an evaporation-cooled 1,000 megawatt electric power plant would consume between 310 and 680 acre feet of water per trillion Btus produced (A trillion Btus is approximately equal to the energy contained in 37 thousand tons of bituminous coal, or 153,000 barrels of oil) (See also Table 8.2).

Transportation Policy Issues

In comparing the external effects of alternative ways of exporting energy from coal, the following issues should be considered:

- Water use.
- Air pollution.
- Energy efficiency.
- Flexibility of operation.
- Safety and reliability.
- Employment and tax base.
- Effects on other industries.
- Other related socio-economic and environmental impacts.

EMPLOYMENT, MANPOWER AND TRAINING

Traditionally, coal mining has been a labor-intensive industry requiring relatively large numbers of employees. This has resulted in severe boom-and-bust impacts on employment and on coal mining communities at various times throughout our history. As mining methods have become more mechanized, the need for manpower to produce a specific amount of coal has declined. In fact, coal mine labor has steadily decreased from 1948 to 1969 (Risser and Malhotra, 1976, p. 475) (see Labor Unions and Productivity, Chapter 5). However, since 1970, the number of persons employed in coal mines has been increasing; this increase in manpower needs can be expected to continue into the future.

Manpower needs in the coal industry can be divided into two major categories--underground mines and surface mines. Katell et al. (1976a and 1976b) has estimated manpower requirements and labor costs for each type of mine and the results are summarized in Tables 7.7 and 7.8. Actual jobs and wage requirements are listed in the references.

Table 7.7 lists estimates for various labor production rates. The data shows that generally as the production rate increases, manpower needs increase. However, as the production rate increases from 1.06 to 3.18 million tons per year, the cost of labor per ton of coal produced decreases from $2.76 to $2.35. As the mine size is further increased to 4.99 million tons per year, the cost of labor per ton of coal produced essentially remains constant. These figures indicate the economies of scale of underground coal mining labor costs, with constant returns to scale appearing at a production rate of 3.18 million tons. This means there are dollars which can be saved per ton of coal produced as the mine increases production to 3.18 million tons. (There is an economic incentive to hire additional manpower and expand production from smaller production rates.) But this saving disappears as that production figure is expanded above 3.18 million tons. (The incentive to expand and hire additional employees also disappears.)

A summary of the estimated manpower and labor costs for surface coal mines is presented in Table 7.8. The rates of coal production vary according to possible locations of the mine. The table indicates that as the production rate of coal increases, manpower requirements rise while the cost of labor per ton of coal produced decreases. These production figures suggest there are economies of scale in surface mining so there may be an economic incentive to expand production.

However, these economies of scale might also have arisen only from the location of the mine (e.g., strip mining labor costs are generally smaller in the Northern Great Plains as compared to the interior or eastern mines, no matter what the production rate). By comparing the two tables, it can be seen that the surface mines are much more productive, require a much smaller labor force, and cost per ton of coal is smaller than in underground mines.

With anticipated increases in coal production in the United States, there may be problems in finding enough qualified workers, at all levels of skills and abilities, to meet anticipated demand. This could be a very serious constraint on

TABLE 7.7

MANPOWER AND LABOR COSTS FOR UNDERGROUND COAL MINES (1975 dollars)

	Million tons per year			
	1.06	2.04	3.18	4.99
Total labor and supervision	246	428	633	997
Cost of labor and supervision	$2,928,900	$5,049,700	$7,464,100	$11,685,000
Cost per ton of coal produced	$2.76	$2.48	$2.35	$2.34

SOURCE: Katell et al., 1976b, pp. 8, 17, 26, 35.

TABLE 7.8

MANPOWER AND LABOR COSTS FOR SURFACE COAL MINES (1975 dollars)

	Million tons per year and location		
	4.8 Eastern	6.72 Interior	9.2 Northern Great Plains
Total labor and supervision	178	184	213
Cost of labor and supervision	$2,308,600	$2,383,500	$2,629,400
Cost per ton of production	$.48	$.35	$.29

SOURCE: Katell, et al., 1976a, pp. 10, 19, 28.

coal production; efforts should begin immediately to provide for an adequate supply of labor for the future if coal production is to increase.

As an example, mining technicians possessing some of the abilities of mining engineers are needed. Three Colorado institutions (Trinidad State Junior College, Delta-Montrose Vocational Technical School and Colorado Northwest Community College) have ongoing training programs in mining technology (including both safety and basic mining technology training) with high levels of industry participation (Hovis, 1978, p. 9). Table 7.9 shows employment in Colorado coal mines for the period 1960-1977. It reflects an increasing trend since the early 1970s which can be expected to continue in the future. Additionally, more manpower will be needed for coal electricity generating and coal conversion facilities.

SOCIO-ECONOMIC IMPACTS

The impact upon communities located near energy-related developments has been extensive. It is often difficult to prevent or reduce the adverse impacts since each community has its own specific problems and remedies for solution of these problems. However, certain negative impacts do tend to be repeated in most of the boom-town or energy-affected communities. These impacts from in-migration (where necessary to fill new jobs) include lack of adequate housing, recreation facilities, municipal services (fire, police, medical, education, etc.) and infrastructure combined with increases in population, crime rates, mental health problems, traffic and noise. Some of these impacts can be anticipated now, but problems still arise because there is not adequate funding in many cases to effectively decrease these impacts.

On the positive side, impacts from energy development yield greater job opportunities, increased incomes, and a wider selection of items for purchase. However, in most energy impacted communities, the adverse impacts presently appear to outweigh the positive impacts, primarily because of the lack of coordinated planning and the capital needed to maintain the quality of life which existed before development. Research efforts are now underway which attempt to assess both positive and negative impacts.

A Federal Energy Administration (FEA) report (1977, pp. 36-100) details some of the areas of actual or prospective impact problems in Colorado. The greatest percentage of growth in population from 1970 to 1977 had taken place in the towns of Oak Creek (164%), DeBeque (158%), Carbondale (120%), Silt (96%), and Grand Valley (85%). From the population projections it would appear that the greatest growth (over 200 percent from 1970 to 1980) will probably occur in the communities of DeBeque (545%), Grand Valley (474%), Collbran (344%), Oak Creek (315%), Dinosaur (305%), Rifle (300%), Meeker (288%), Carbondale (286%), Hayden (285%), Steamboat Springs (285%), Rangely (283%), and

TABLE 7.9

EMPLOYMENT IN COAL MINES IN COLORADO

Year	Employment
1960	2051
1961	1657
1962	1594
1963	1393
1964	1474
1965	1500
1966	1518
1967	1381
1968	1364
1969	1357
1970	1385
1971	1389
1972	1361
1973	1534
1974	1736
1975	1975
1976	2259
1977	2944

SOURCE: Colorado Division of Mines Annual Coal Statistics, 1960-1977.

Silt (200%). Most of these increases reflect coal and/or oil shale development. This increased population is primarily on the Western Slope of Colorado in the counties of Garfield, Moffat, Routt, Rio Blanco, and Delta.

The FEA report further states that eleven of the thirty-eight impacted communities in Colorado had not established building codes and sixteen had no zoning ordinances; however, eleven communities had planning personnel on their administrative staff; and nineteen had official planning documents for use by local officials. Along with building codes and zoning ordinances to assist in managing growth (and land use management), twenty-one of the thirty-eight communities had instituted subdivision regulations. From these facts, it appears that attention needs to be directed toward securing additional professional personnel in many of the communities, along with encouraging more of the concerned citizens to participate in local planning.

Colorado communities impacted by coal development, according to a priority ranking by the FEA, based on input from local, regional, and State level sources, are listed in Table 7.10. Communities within "A" ranking should be given the highest priority in deciding where government aid should be directed to help alleviate impact. Those communities with a "B" ranking should be given second priority, and so forth. However, category "D" is not necessarily a fourth priority category. While the "D" communities have potential to be included in the first three categories, they are not judged to be experiencing significant coal-related growth now. The "E" grouping does not represent a fifth priority, but is a category including loosely defined communities which require basic infrastructure investments before they can be formed into viable communities. Category "F" includes those communities impacted by energy development outside of the State.

It must be emphasized that this categorical listing is at least partly subjective (formed from the opinion of the Socio-Economic Impact Coordinator's Office) and may not be wholly appropriate. Cities such as Craig and Carbondale are entirely non-comparable because of site specific problems (Gilmore, 1978). However, the listing does give decision-makers some idea of the relative need for aid (monetary or otherwise) among communities. This list cannot be used to specify the types of problems occurring in each site; that must be done only after careful observation and communication. In addition to these community impacts, certain regional impacts such as transportation (discussed earlier in this chapter) must be considered.

Financial assistance to lessen adverse socio-economic impacts can come from State or Federal government, or industry. The Federal government has designed certain aid and information-gathering procedures to facilitate this service. For example, the Department of Energy maintains a Socio-Economic Impact office which has compiled much useful information. Unfortunately, in many cases, Environmental Impact Statements prepared by the federal departments have been inadequate in assessing socio-economic impacts from proposed developments. Additionally, many federal aid programs are not geared to help these communities at the time when adverse impacts are most severe, nor do they always fit the characteristics of the community (such as aid programs for high unemployment areas, which are normally not problems in these areas).

The role of industry in helping to mitigate adverse community impacts varies among companies. Some companies demonstrate social responsibility by conducting orientation programs for employees, hiring local inhabitants extensively before bringing in outsiders, arranging for housing, financing local

TABLE 7.10

RANKING OF COLORADO COAL PRODUCTION OR CONVERSION IMPACTED COMMUNITIES

Category A

Communities experiencing significant increases in population due to build-up of the coal industry. Needs are immediate.

Craig, Moffat County
Hayden, Routt
Carbondale, Garfield

Category B

Communities where the sources of impact are the most imminent, and the resources to deal with those impacts are the most limited.

Paonia, Delta County
Hotchkiss, Delta
Somerset, Gunnison
Rangley, Rio Blanco*
Meeker, Rio Blanco*
Oak Creek, Routt

Category C

Communities which have a high potential for significant growth, the degree of impact being dependent upon unstable economic conditions. Needs must be anticipated to avoid future crises.

Glenwood Springs, Garfield County*
DeBeque, Mesa*
Cedaredge, Delta
Orchard City, Delta
Delta, Delta
Walden, Jackson
Yampa, Routt
Steamboat Springs, Routt

Category D

Communities not now experiencing significant energy related growth but which require continued monitoring due to potential activity. Includes areas where large quantities of coal or where industry had indicated intentions of certain activity.

Collbran, Mesa County*
Palisade, Mesa*
Fruita, Mesa*
Dinosaur, Moffat*

TABLE 7.10 (continued)

Trinidad, Las Animas
Walsenburg, Huerfano
LaVeta, Huerfano
Aguilar, Las Animas
Durango, La Plata
Crawford, Delta
Bennett, Adams

Category E

Unincorporated areas, or other areas basically rural in nature, which are now or could be affected by coal development, and need assistance if they are to become viable communities.

Somerset, Gunnison County
Cokedale, Las Animas
Redstone, Pitkin
Phippsburg, Routt
Watkins, Adams
Maybell, Moffat

Category F

Communities which are experiencing growth resulting from development outside the State of Colorado.

Cortez, Montezuma
Pagosa Spring, Archuleta

SOURCE: Federal Energy Administration, 1977, pp. 48-50.

*Also impacted from oil shale development in addition to coal.

Note: This priority list is only temporary in nature, and communities may periodically move from one category to another or even off the list as conditions change.

planning efforts, and engaging in other positive actions.

Some assistance has come from State government. One example is the Colorado Energy Impact Assistance Plan, which evolved in part from the recommendations included in the Boom Town Financing Study (Bickert, Browne, Coddington and Associates, 1976; and Bolt, Luna and Watkins, 1976), the activities of the Governor's Socio-Economic Impact Office and actions of the 1977 Colorado General Assembly. Special adverse impact funds under this plan are limited, but are intended to fill the gap when other sources are not available to support on-going aid programs.

Among the major State sources for providing local financial assistance are: royalty payments from the Federal Mineral Lands Leasing Act of 1920 (through the Local Government Mineral Impact Fund), interest earned on the Oil Shale Trust Fund, the State Severance Tax, and on-going State and federal financial assistance programs. Even with these sources it is highly unlikely that available money will be adequate to meet needs. Even so, there are numerous State technical assistance programs available to local communities. These include State agency programs, Local Impact Assistance Teams, council of government teams, university and other educational institutions, industry, and federal agency programs (Colorado Socio-Economic Impact Office, 1978, pp. i-vi).

The United States Congress is currently considering national energy impact legislation (S. 1493, as amended), sponsored by Senator Gary Hart (D-Colo). If passed in its present form, this bill would provide financial help, especially loans and loan guarantees, in the early stages of energy development when other sources of revenue are not available. The bill would also provide technical assistance to energy impacted communities. Possible action on this legislation is expected sometime in the near future.

MINING SAFETY

Although coal mining is historically a hazardous occupation, much progress has been made in reducing accidents in both surface and underground mining. Surface mining has safety problems involving the use of heavy equipment and explosives. Underground mining safety problems which merit serious concern are ventilation, methane concentration and coal dust control, roof support, and fire and explosive control.

Progress in safety is evident from the data compiled by the Mine Safety and Health Administration (Allen, 1978) listing fatalities in underground operations from 1935 to 1977. A reduction from 1.5 to 0.36 fatalities per million man-hours worked was recorded over this time period. However, the tabulation did not include disabling injuries, such as black lung disease. Fatalities and injuries should also be related to production. This may be more meaningful in terms of real variations in production per man-hour.

Based on work injuries statistics, subsurface coal mining is 1.5 times more hazardous than work in the construction industry, and 3 times more hazardous than work in manufacturing. Subsurface coal mining has eight times the national average incidence of respiratory diseases, with a related death rate five times the industrial average. Table 7.11 shows Colorado accident data for the period 1960-1977.

The primary federal law regarding coal mining health and safety is the Federal Coal Mine Health and Safety Act (CMHSA) of 1969. The legislation and regulations were originally to be administered by the Department of the Interior through the

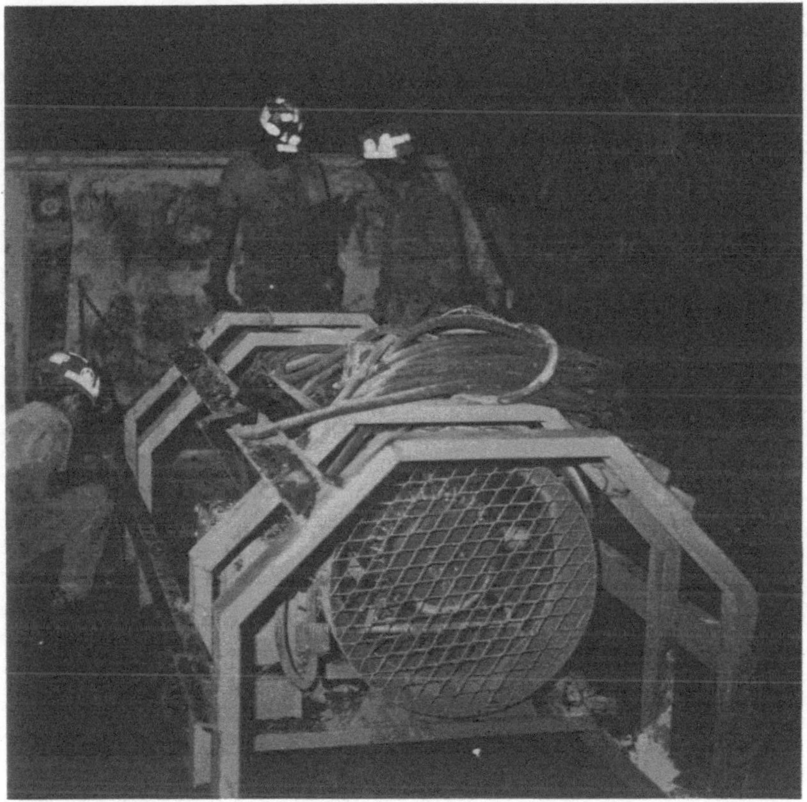

Plate 7.2 Maintenance crew moving auxiliary fan to new location. These fans are equipped with noise attenuators and safety frames and are used to direct ventilation air into specific locations such as a working face during mining, to keep a sufficient volume and velocity of air moving through these areas. (Courtesy Colorado Westmoreland Coal Co.)

TABLE 7.11

COLORADO COAL ACCIDENTS

Year	Fatal Accidents	Nonfatal Accidents
1960	2	78
1961	5	76
1962	3	74
1963	1	70
1964	3	67
1965	11	55
1966	4	62
1967	2	38
1968	6	53
1969	4	38
1970	0	56
1971	1	67
1972	5	63
1973	3	218
1974	2	306
1975	6	285
1976	3	251
1977	2	246

SOURCE: Colorado Division of Mines Annual coal statistics, 1960-1977.

Mining Enforcement and Safety Administration (MESA), but responsibility was shifted to the Department of Labor through the Mine Safety and Health Act Amendments of 1977. The Colorado mining laws, which predate the CMHSA, are quite comprehensive, with at least the same degree of stringency as the federal standards (U.S. Department of the Interior, 1976, O. VI-86). Both Federal and State laws call for strict mining engineering regulations (distances between crosscuts, roof support parameters, etc.), and for frequent inspections of coal mining operations.

One of the major impacts of the Mine Safety and Health Amendments Act has been the implementation of training programs.

Plate 7.3 Mike Burum, mine safety staff, aboard the left boom of a twin-boom roof bolter, used to drill holes in the roof and then insert and spin the resin bolts. This equipment was designed by and built for the Orchard Valley mine. It carries its own jack assembly and temporary roof supports, precluding the necessity of setting safety jacks most of the time. (Courtesy Colorado Westmoreland Coal Co.)

Coal mine operators, at their own expense and during normal working hours, must provide at least forty hours of safety instruction for each new underground miner and twenty-four hours for new surface miners. The act also requires eight hours of annual retraining for all miners, as well as retraining for any new miner assigned to a new task (Mining Engineering, 1978, p. 514).

8
Technological Factors Affecting Coal Use

POWER GENERATION

Energy Conversion Efficiencies

Public utilities are the major users of coal in the United States. Consequently, the efficiency with which the fuel is burned to generate steam is a major determinant of how much coal is consumed for electrical power generation. Since the efficiencies of modern power plants have only been as high as 40 percent, as much as 60 percent of the energy input is lost as waste heat. In addition, state-of-the-art power plants now contain various air pollution control equipment, which further reduces their total efficiency to about 37 percent, as indicated in Table 8.1.

As an example of the fuel savings possible from better operating efficiencies, consider a 1,000 MW power plant which uses some 3 million tons of coal per year. A 5 percent increase in power plant efficiency translates into 15,000 tons of coal per year saved per power plant. Many technologies for increased improvements in combustion are under development. However, in the final analysis, efficiency gains will have to be weighed against capital costs, reduction in stack emissions, and other constraints.

Brief note is made here of one of more dramatic possibilities for improved efficiencies. The improved system is called magnetohydrodynamics (MHD). This is a process where coal is burned at extremely high temperatures (on the order of 5,000° F). Injection of small quantities of salt into the airstream forms an ionized gas or plasma (a gas whose particles carry a small electric charge). The plasma is then forced at high speed through a channel surrounded by an intense magnetic field. This action converts the gas movement directly into electricity, in a system with few mechanical moving parts.

The hot gas is then recycled through the process; but prior to its reuse, its heat (at still high temperatures) can be

TABLE 8.1

COAL CONVERSION EFFICIENCIES AND PRODUCT PRICES

Process	Efficiency, %	Product Price $/million Btu
Electricity Generation		
Conventional Steam Cycle	37	(1975 $) 10.26 - 12.60
Combined Cycle (See Glossary)	41 - 47	9.08 - 11.13 (1975 $)
Liquid Fuels		
First Generation (Fischer-Tropsch)	56	(1977 $) 5.44
Second Generation (SRC - II)	68*	(1978 $) 3.33 - 5.83
Solid Fuels		
Second Generation (SRC - I)	78	3.00 - 5.25
SNG From Coal		
First Generation (Lurgi)	59	(1976 $) 3.22
Second Generation (HYGAS) (See Glossary)	78	(1976 $) 2.71

SOURCE: Energy Topics, 1978, p. 2.

*Estimated

transferred to water, converting the water to steam which, in turn, is used to drive a conventional steam turbine--creating additional electricity. Theoretical efficiencies for MHD are estimated at 75 percent, or twice that of today's coal-fired generating plants. Potential coal consumption savings using MHD are also dramatic, cutting overall coal need by 50 percent. However, while MHD's potential is enormous, technology development may require at least another decade before a prototype plant can be built and test-operated.

Water Use

Virtually all the water required by a coal-fired electric generating plant is used by its condenser cooling system. Since about two-thirds of the plant's energy input is rejected as waste heat, this heat must be removed by water (whose return temperature to its original source is limited by technological and environmental constraints). This heat transfer function requires large amounts of cooling water in each plant.

There are several basic types of cooling systems for power plants being used: once-through processing, cooling ponds, wet towers and dry towers (See Glossary). Anticipated environmental and water quality legislation may impose zero discharge requirements on energy-related projects in the arid and semi-arid West. Cooling technology in the Rocky Mountain region is expected to be confined to wet or dry towers, or hybrids of these systems (Los Alamos Scientific Laboratory, 1978, p. 43).

It is noteworthy that dry tower cooling requires less than 1 percent of the water used in wet towers, but severe economic penalties inhibit its use. (See also Power Plant chapter for relative quantities of water needed for each type of cooling system.)

TYPICAL COAL-FIRED POWER PLANT

Some of the typical characteristics of a 1,000 MW coal-fired plant are shown in Table 8.2. They include: water use, land requirements along with material, manpower, and fuel needs.

The coal-fired electric generating plant consists of a series of subsystems which include a furnace, steam generator (boiler), and a steam-driven turbine-generator. Steam that is condensed continuously at the turbine exhaust is pumped back to the boiler as feedwater.

Coal, before it can be used to generate power, must first be pulverized to the consistency of talcum powder and then fed into the furnace pneumatically through nozzles.

The thermal efficiency of the combustion conversion process is often expressed as the net-plant-heat-rate (NPHR) or the amount of fuel heat required to generate and deliver one kilowatt-hour of electricity to the transmission line. For modern coal-fired plants, NPHR ranges about 10,455 Btu/Kw-hr. Thermal efficiency is calculated by dividing 3,413 (the standard Btu equivalent of 1 Kw-hr) by NPHR and multiplying the result by 100 for percentage value. Thus modern plants have efficiencies of about 36 percent.

The elimination, or rejection of waste heat is necessary to generate electricity. A 1,000 MWe* plant must reject 4.2 billion Btu/hr, over and above the 1.1 billion Btu/hr heat lost

*MWe refers to megawatts of electrical production.

TABLE 8.2

CHARACTERISTICS OF A 1,000 MW COAL FIRED PLANT

1. Plant Water Use for Various Cooling Systems

Cooling System	Total Water Consumption (acre-feet/yr)
Cooling Pond (a)	9,700
Spray Pond (b)	10,800
Evaporative Tower	8,800
Dry Cooling	Minimal

(a) Surface area required = 1000 acres
(b) Surface area required = 50 acres

2. Other Factors

1) Land requirements:	Plant	10 acres
	Cooling pond	1000 acres
	Ash pond	100 - 200 acres
	Wet-scrubber slush pond	250 acres
	Misc.	200 acres
2) Material requirements:	Concrete	67,000 tons
	Steel	48,100 tons
	Misc.	1,000 tons
3) Manpower requirements:	Design	633,000 man-hours
	Construction	6,000,000 man-hours
	Suprv. & Admin.	900,000 man-hours
	(early build-up to 500-600 people; peaks in 3rd year at 1200-1400; gradual reduction; time span - 7 years, minimum)	
	Plant operation: 160 people	
4) Fuel requirements:	3 million tons of coal per year.	
5) Wastes Sludge:	900 thousand tons per year from scrubbers (if coal contains 3.5% sulfur)	
6) Ash:	300 thousand tons per year (10% ash)	

SOURCE: Resource and Land Investigations, 1975; Byron and Saleem, 1978, p. 147.

from the boiler in the form of hot stack gases. Removal of this excess heat is accomplished by wet or dry cooling techniques. Dry cooling techniques, although more expensive, eliminate the consumptive use of water thus enabling more flexibility in plant location if water supplies are a major constraint. A comparison of the water consumption and other factors of various cooling systems is also given in Table 8.2.

COAL GASIFICATION AND LIQUEFACTION

Among the major alternatives to the direct use of coal are processes which convert the solid fuel to transportable gas and liquid forms, e.g., synthetic natural gas (SNG) with varying Btu values up to those as high as natural gas, gasoline, heavy liquid boiler fuel and methanol.

Gasification

The development of commercially viable high-Btu coal gasification processes is a major component of the U.S. synthetic fuels program. High Btu gas in this case refers to "pipeline quality" SNG, whose caloric content is about 1,000 Btu/standard cubic feet (scf), equivalent to that of natural gas. Fixed bed and fluidized bed systems, of which Lurgi and HYGAS are respective examples, offer the greatest promise for early commercialization (Stringer, 1978).

The equipment for a gasifier consists of a vessel within which a mixture of oxygen and steam is channeled through a bed of burning crushed coal. In its simplest form, coal is fed in at the top of the vessel while an oxygen/steam mixture enters at the bottom. With combustion taking place at the bottom, the hot gases rise toward the top, and accumulated heat is removed just above the burning zone. This permits the heat to be recovered for an increase in processing efficiency. The so-called off-gas, removed at the top, is a mixture of hydrogen, carbon monoxide, water vapor and coal volatiles. Noncombustibles are removed at the bottom of the gasifier.

The gas produced at this juncture has low Btu value, and upgrading to pipeline quality requires two additional major steps: first, carbon monoxide is combined with water forming carbon dioxide and hydrogen. The hydrogen produced from this last process is reacted with carbon monoxide to yield methane (the major component of natural gas) and water. This end step is called methanation. In the second major step, the gas is dried; after that, sulfur is often removed in the form of hydrogen sulfide.

Water consumption in an SNG facility amounts to about 10-15 percent of the total water used; the remaining non-consumptive water is used to cool pumps and the gas itself. Cooling water sources will be derived from treated process water. Total actual water consumption is process dependent, but is estimated to range from 4,000 acre-feet per year (arid sites) to 12,000 acre-feet per year (water plentiful sites) for a plant capacity of 100 million standard cubic feet/day (Los Alamos Scientific Laboratory, 1978).

As an alternative to burning solid coal fuels, coal may be gasified and burned directly in a boiler to generate steam, or in a gas turbine as part of a combined cycle, thus generating electricity without engaging the second and third steps. Production and transportation economics in this case, and in terms of Btu values, permit transport of the gas to major users only short distances from the gasifier.

These economic constraints led DOE's Laramie Energy Research Center to conduct tests over the past several years on in-situ or underground coal gasification at Hanna, Wyoming. The process involves the drilling of boreholes in a section of the coal deposit, rubblizing the coal subsurface using explosives, and, following that, igniting the coal and forcing air (or oxygen and steam) into the burning zone. The combustible gases (carbon monoxide, hydrogen and methane), are collected at the surface through another borehole. This process is continued until the fractured seam section is no longer productive. The resulting gases, from using air alone have a value of about 150 Btu/scf.

Further research performed by the Lawrence Livermore Laboratory (University of California) near Gillette, Wyoming, using oxygen-steam mixtures, demonstrated an increase in the heat content of the gas to 265 - 300 Btu/scf.

The in-situ process offers four potential advantages over surface gasification processes: (1) smaller capital investment may lead to lower product costs; (2) environmental damage may be far less than either surface or underground mining; (3) the safety hazards to miners are reduced; and (4) coal resources too deep, too thin, or too poor in quality for economic mining may be used.

Other than the possible effects on ground water now being investigated, the in-situ process appears to create a minimal environmental impact. The excellent results of early research suggest that the process may be an alternative not only to surface gasification, but also to mining and movement of solid energy fuels. A comparison of gas yields from in-situ testing with those of the Lurgi Process are shown in Table 8.3.

Liquefaction

This process involves increasing the hydrogen-carbon ratio (H/C ratio) in coal. The higher the ratio (it is highest in gasoline), the lower the boiling and combustion points of the product. Liquefaction processes are classed as direct and indirect, with the latter actually an extension of the gasification process. In indirect liquefaction, the gas containing carbon monoxide and hydrogen is reacted with a catalyst to produce a wide variety of hydrocarbons ranging from methane to light oils. The H/C ratio can be varied, but the process

TABLE 8.3

TYPICAL GAS YIELDS (WATER-FREE) IN PERCENT FROM IN-SITU (UCG) AND CONVENTIONAL GASIFIERS

	UCG*	Lurgi
Hydrogen	18.6	23.1
Methane	3.6	4.1
Nitrogen and argon	47.5	42.1
Carbon monoxide	16.5	15.1
Carbon dioxide	13.1	15.1
Hydrogen sulfide	0.1	0.1
Ethane +	0.6	0.6
Higher heating value, Btu/SCF	161.0	175.0

SOURCE: Edgar, T. F., 1977, p. 79.

*Hanna, Wyoming data DOE

is rather expensive as experience has shown in the operation of the Sasol plant in South Africa.

Direct coal liquefaction, on the other hand, uses a hydrogenation/desulfurizing technique, which is a process of adding hydrogen to, and removing sulfur from, coal. This is accomplished with rapid, turbulent hydrogen flow moving a coal slurry through a catalyst bed. Recycled product oil is used as the slurry base. Water consumption is estimated at 200 acre-feet per year for a 1,000 bbl/day liquefaction plant. The Synthoil process shows promise of early commercialization of the direct coal liquefaction (Los Alamos Scientific Laboratories, 1978).

Although there is no commercially proven liquefaction technology at present, several processes, producing yields up to three barrels of oil per ton of coal are now in large-scale development. They include the Exxon Donor Solvent and the HRI-H-Coal processes. Small-scale research and development on other processes is being done by Pittsburgh Energy Research Center (DOE), Dow Chemical, Conoco, Lummus, and others (Swabb, 1978).

Methanol

Through suitable processing steps, coal gasification can be modified to produce methyl alcohol (methanol) as an end product. Mobil Oil Corporation recently demonstrated a process extending methanol to synthetic gasoline in pilot plant testing. It is too early to contrast this process route to liquid fuels with those based on liquefaction, but the research does indicate an additional option for converting coal to liquid fuels (Chemical and Engineering News, 1978).

9
Coal Production Constraints

There exist various factors which affect coal production today. These factors center around uncertainties in economics, technology, environmental concerns, socio-economic issues, and governmental regulation. Until these issues are at least partially resolved, coal production increases in the future will be unpredictable.

ECONOMIC FACTORS

Adequate coal reserves exist today, but it is uncertain how much and what kinds of coal will be most desirable in the market in the future. Capital and operating costs have escalated during the past years, and with the high risk involved in these large facilities, financing may prove difficult. Other uncertainties include unions and productivity, effects of taxes, and environmental and reclamation requirements and costs.

TECHNOLOGICAL FACTORS

Coal exploration, recovery and beneficiation techniques are well known today, but there still exists a need for productivity increases and increased efficiency through technology. To a certain degree, there is still inadequate geologic knowledge of coal resources. It is uncertain if there will be enough unskilled and skilled manpower necessary for large industry expansion. Gasification, liquefaction, and fluidized bed processes hold high promise of coal utilization in the future; however, they are largely untested today and much research needs to be done.

ENVIRONMENTAL FACTORS

Environmental regulation has been a major constraint for coal production since major legislation was enacted in the early 1970s. Sulfur removal and reclamation needs appear to be the most troublesome issues to industry. Other uncertainties include safety concerns, water needs and availability, and transportation methods which are efficient yet environmentally sound.

SOCIO-ECONOMIC FACTORS

Increased coal production will certainly result in large population increases in sparsely populated areas, both from basic mine personnel and indirect (service) employment. Resulting problems can often be anticipated, but a lack of funding at the necessary times prevents significant action. Planning and determination of where responsibility lies are important needs in these situations.

REGULATORY FACTORS

Besides the environmental constraints, unavailability of federal leases in recent years and uncertainties of future federal leasing procedures have severely hampered industry planning. The permitting procedure is time consuming, costly, complex, confusing, and requires duplication of effort at different governmental levels.

10 The Future: Observations and Concerns

Given coal's abundance, widespread distribution, relative ease of access for extraction, chemical versatility and need, increases in the use of coal are readily apparent. The extent of this increase may be accelerated by administrative pressure and resulting legislation to convert oil and gas-fired power plants to coal, and by an acceleration in technological development of coal gasification and liquefaction processes. On the other hand, use may be moderated by the development of practical alternative energy sources, stringent environmental constraints on mining, transportation and combustion--our ability within the State to mine, move and burn coal--along with related socio-economic impacts.

In addition to large capital requirements, the time needed to increase coal production creates a difficult hurdle to surmount. As an example, it requires five to nine years to bring a new underground coal mine into production and from one to twelve years for a surface mine. Orders for capital equipment for either mining operation face a backlog of some five years. In addition, available manpower, hardware for conversion of oil and gas burners to coal and installation of environmental control equipment will also limit the speedup of coal production. Nonetheless, coal production in Colorado for 1985 is forecast to triple the 1977 output of 11.9 million tons.

One national viewpoint about the future of coal development, given by Mr. Carl E. Bagge, president of the National Coal Association, was quoted in Chemical and Engineering News (1977b):

> The coal industry is confident that it can meet the demand for its product in the next decade. We have an ample supply of problems, but we can handle the ones where the solution is in our hands. We can recruit more miners and train them sufficiently. We can find the capital, if something extraneous does not interfere.

> The problems that really concern us lie outside our grasp, beyond our capacity to manage them. These are the problems of government policy. . .
>
> I do not believe that energy and environmental concerns are always necessarily in conflict, nor that energy must always win out if such a conflict occurs, but I do believe that we should look at specific cases as well as lofty objectives. . . .

Other concerns for the future of coal development reflect the need for improvements in technology to reduce capital costs for hardware used in power generation and conversion efficiencies and the performance of pollution control systems. We may eventually see the emergence of energy parks, industrial complexes in remote areas which are dedicated to power generation, plus the production of chemical feedstock through coal gasification or liquefaction processes. All these factors could represent an expansion of the industrial base on a preplanned basis using the state's large coal resources.

Solutions of Colorado's (and the nation's) coal production, transportation and end-use problems require a level of integration and interrelationship, achievable with very great difficulty and time. And time, unfortunately, is not on our side.

References: Coal

Adams, H., 1976. Objectives and organization of coal exploration projects, in The First International Exploration Symposium, London, May 1976, Proceedings: Miller Freeman Publications.

Allen, Betty, 1978. Personal communication: Mine Safety and Health Administration, July 13, 1978.

American Gas Association, 1977. A comparison of coal use for gasification vs. electrification: Arlington, Va.

Averitt, P., 1975. Coal resources of the United States: U.S. Geological Survey Bulletin 1412, January 1, 1974.

Bertoldi, M. J., 1977. Preliminary economics of mining a thick coal seam by dragline, shovel-truck and scraper mining systems: U.S. Bureau of Mines Information Circular 8761.

Bickert, Browne, Coddington, and Associates, 1976. Boom town financing study, vol. 2: Colorado Department of Local Affairs, Denver, Colorado.

Bolt, R. M., Luna, D., and Watkins, L. A., 1976. Boom town financing study, vol. 1: Colorado Department of Local Affairs, Denver, Colorado.

Byron, R. A., and Saleem, A., 1978. Particulate and sulfur oxide control options for conventional coal combustion, in The Fifth Energy Technology Conference, April 1978, Washington, D.C., Proceedings: Government Institutes, Inc., R. I. Hill, Ed.

Chemical and Engineering News, 1977a, Economics a plus for coal slurry pipelines: June 27, v. 55, no. 26.

Chemical and Engineering News, 1977b, Coal: one view of its problems: September 12, v. 55, no. 37.

Cheronis, N. P., 1977. Advanced mining systems emerging-triggered by heavy federal funding; in Coal Age Operation Handbook of Underground Mining: New York, McGraw-Hill, Inc., N. P. Cheronis, Ed.

Colorado Department of Natural Resources, 1978. Letter to director of U.S. Bureau of Land Management: February 14, 1978.

Colorado Division of Mines, 1960-1977. Annual coal statistics: Denver, Colorado.

Colorado Division of Mines, 1978. Monthly coal production report: Denver, Colorado, May 24, 1978.

Colorado Mining Association, 1977. State of Colorado severance tax; in 1977 Mining Year Book: Denver, Colorado.

Colorado Socio-Economic Impact Office, 1978. Colorado energy impact assistance plan; Denver, Colorado, January, 1978.

Cornette, A. J., 1976. Ten year outlook in U.S. coal mining, in 1976 Mining Year Book: Denver, Colorado, Colorado Mining Association.

Edgar, T. F., 1977. Technical, economic and environmental evaluation in in-situ coal gasification: In Situ, v. 1, no. 1.

Energy Topics, 1978: Chicago, Institute of Gas Technology, June 19, 1978.

Energy Technology Handbook, 1977. McGraw-Hill, Inc., D. M. Considine, Ed.

Environmental Science and Technology, 1976. How the coal slurry pipeline in Arizona is working?: November, v. 10, no. 12.

Federal Energy Administration, 1976a. Coal, the other resource: Washington, D. C., GPO 906-949.

Federal Energy Administration, 1976b. Coal mine expansion survey, Washington, D.C., Contract No. CO-05-50313-00.

Federal Energy Administration, 1976c. Western coal workshops, Chicago, summary of proceedings: FEA/G-76/437.

Federal Energy Administration, 1977. A report--regional profile of energy impacted communities: July, 1977.

Gage, S. J., 1977. Control technology bridges to the future: Energy/Environment II, U.S. Environmental Protection Agency, EPA-600/9-77-012.

Gibson, L., 1978. Do Coal and Water Mix?: Colorado Business, Denver, Colorado, February, v. 5, no. 2.

Gilmore, J., 1978. Personal Communication, September 6, 1978.

Harrison, W. G., 1978. Coal-based electricity and air pollution control - a case of solvent refined coal, in The Fifth Energy Technology Conference, April 1978, Washington, D.C., Proceedings: Government Institutes, Inc., R. I. Hill, Ed.

Hasbrouck, W. P., and Hadsell, F. A., 1978. Geophysical techniques for coal exploration and development, in The Second Symposium on the Geology of Rocky Mountain Coal, Colorado Geological Survey, Proceedings: H. E. Hodgson, Ed.

Hornbaker, A. L., Holt, R. D., and Murray, D. K., 1976. 1975 Summary of coal resources in Colorado: Colorado Geological Survey Special Publication Number 9, Denver, Colorado.

Hovis, H., 1978. Education and training, a summary of mineral industry activities in Colorado - 1977, Part 1, Coal: Colorado Division of Mines, Denver, Colorado.

Hunter, D. W., 1977. Trends in coal mining systems - conventional, continuous, longwall, shortwall, in Coal Age Operating Handbook of Underground Mining: McGraw-Hill, Inc., N. P. Cheronis, Ed.

Jackson, D. M., and Schmid, B. K., 1978. SRC-II review of development and status, in The Fifth Energy Technology Conference, April 1978, Washington, D.C., Proceedings: Government Institutes, Inc., R. I. Hill, Ed.

Jacobsen, P. S., 1978. Coal preparation--computer simulation of plant performance: Mineral Industries Bulletin, Colorado School of Mines/Colorado School of Mines Research Institute, Golden, Colorado, January, v. 21, no. 1.

Jones, J. R., 1977. The process of developing a western coal mine, in 1977 Mining Year Book: Colorado Mining Association, Denver, Colorado.

Katell, S., and Hemingway, E. L., 1974. Basic estimated capital investment and operating costs for underground bituminous coal mines: U.S. Bureau of Mines Information Circular 8632, Washington, D.C.

Katell, S., Hemingway, E. L., and Berkshire, L. H., 1976a. Basic estimated capital investment and operating costs for coal strip mines: U.S. Bureau of Mines Information Circular 8703, Washington, D.C.

Katell, S., Hemingway, E. L., and Berkshire, L. H., 1976b. Basic estimated capital investment and operating costs for underground bituminous coal mines: U.S. Bureau of Mines Information Circular 8682A, Washington, D.C.

Kinney, P. W., 1977. Environmental permit directory: Office of the Richard D. Lamm, Governor of Colorado, Denver, Colorado.

Lees, A., 1978. Personal communication, August 23, 1978.

Los Alamos Scientific Laboratory, 1978. Draft on the impacts of the national coal utilization assessment of the Rocky Mountain region, S-2:78:2.

Lowrie, R. L., 1977. Western coal in the U.S. energy picture, in 1977 Mining Year Book: Colorado Mining Association, Denver, Colorado.

Lusk, B. E., 1978. What's in the surface mining rules?: Coal Age, February, v. 83, no. 2.

Martin, J., 1978. Unpublished background paper for Federal coal leasing: Presented to Denver Coal Club, Denver, Colorado, May 9, 1978.

Merritt, P. C., and Davis, H., 1977. Longwall mining: avenue to safety, productivity and resource recovery, in Coal Age Operating Handbook of Underground Mining, New York, McGraw-Hill, Inc., N. P. Cheronis, Ed.

Mining Engineering, 1978. Coal--disappointment amid high hopes: May, v. 30, no. 5.

Mining Record, 1978. Longwall mining can produce 10 times more coal per miner: March 15, 1978.

Mountain Plains Federal Regional Council, 1977. The permitting process for surface mining federal coal with recommendations for improvement, Denver, Colorado.

Murray, D. K., 1977. Colorado coal, a versatile resource: presented to The Annual Meeting of the South Central Section, Geological Society of America, El Paso, Texas, March 17, 1977.

Murray, D. K., 1978a. Background paper for Colorado's coal resources: presented to The RMASA-ASEE Symposium, Colorado School of Mines, Golden, Colorado, April 21, 1978.

Murray, D. K., 1978b. Personal communication, June 7, 1978.

Natural Resources Journal, 1976. Coal taxation in the western states; the need for a regional tax policy: April, v. 15, no. 2.

Office of Technology Assessment, 1978. A technology assessment of coal slurry pipelines: March 1978.

Persse, F. J., Lockard, D. W., and Lindquist, A. E., 1977. Coal surface mining reclamation costs in the Western United States: U.S. Bureau of Mines Information Circular 8737, Washington, D.C.

Plotkin, S. E., 1977. Integrated assessment of energy development in the Western U.S.: Energy/Environment II, U.S. Environmental Protection Agency, EPA-600/9-77-012.

Resource and Land Investigation (RALI) Program, 1975. Prepared for The U.S. Geological Survey by Mitre Corp., MTR-69-88.

Rich, C. H.,Jr., 1978. Projects to expand energy sources in the Western United States: U.S. Bureau of Mines Information Circular 8772, Washington, D.C.

Risser, H. E., and Malhotra, R., 1976. Coal, in Economics of the Mineral Industries, 3rd Edition: American Institute of Mining, Metallurgical, and Petroleum Engineers, New York, 1976.

Speltz, C. N., 1976. Strippable coal resources of Colorado: U.S. Bureau of Mines Information Circular 8713, Washington, D.C.

Spence, H. M., 1976. Mineral severance taxes - causes and effects, in 1976 Mining Year Book: Colorado Mining Association, Denver, Colorado.

Stanwood, R. M., 1978a. Personal communication, July 6, 1978.

Stanwood, R. M., 1978b. A successful energy-environmental tradeoff, the West German coal mining process: presented to Seminar in Energy Economics, Colorado School of Mines, Golden, Colorado, May 1978.

Stermole, F. J., 1974. Economic evaluation and investment decision methods: Investment Evaluations Corp., Golden, Colorado.

Stringer, J., 1978. Materials problems and opportunities in coal conservation processes, in The Fifth Energy Technology Conference, Washington, D.C., April 1978, Proceedings: Government Institutes, Inc., R. I. Hill, Ed.

Swabb, L. E., Jr., 1978. Liquid fuels from coal--from R & D to an industry: Science, February, v. 199, no. 4329.

Taylor, G. C., 1976. Water for western energy resource development (unpublished report): Colorado Energy Research Institute, Golden, Colorado, April 1976.

U.S. Bureau of Mines, 1972. Cost analysis of model strip mining in the United States: U.S. Bureau of Mines Information Circular 8535, Washington, D.C.

U.S. Bureau of Mines, 1976. Mineral facts and problems: U.S. Bureau of Mines Bulletin 667, Washington, D.C.

U.S. Bureau of Mines and U.S. Geological Survey, 1976. Principles of the mineral resources classification system: U.S. Geological Survey Bulletin 1450-A, Washington, D.C.

U.S. Bureau of Mines, 1977. Advanced coal mining technology research, development and demonstration in fiscal year 1977: U.S. Bureau of Mines Information Circular 8730, Washington, D.C.

U.S. Department of Energy, 1977. Inventory of power plants in the United States: Washington, D.C., December, DOE/RA-0001.

U.S. Department of Energy, 1978a. Federal coal leasing and 1985 and 1990 coal production forecasts: Leasing Policy Development Office, Washington, D.C.

U.S. Department of Energy, 1978b. Coal gasification quarterly report: Washington, D.C., May, DOE/ET-0024/4.

U.S. Department of Energy, 1978c. Coal liquefaction quarterly report: Washington, D.C., February, DOE/ET-0026/3.

U.S. Department of the Interior, 1975. Proposed federal coal leasing program environmental impact statement: Washington, D.C.

U.S. Department of the Interior, 1976. Laws and regulations affecting coal: Washington, D.C., Contract 14-01-0001-2115.

U.S. Department of the Interior and Environmental Protection Agency, 1977. Coal cleaning with scrubbing for sulfur control--an engineering economic summary: Washington, D.C., August, EPA-600/9-77-017.

University of Oklahoma, 1975. Energy alternatives: a comparative analysis: Science and Public Policy Program, Norman, Oklahoma.

Wang, F. D., 1978. Personal communication, June 5, 1978.

Weichman, B. E., 1977. Nahcolite use development, in 10th Oil Shale Symposium, Golden, Colorado, Proceedings: Colorado School of Mines, Golden, Colorado, J. H. Gary, Ed.

Yeager, K. E., 1978. Impact of best available control technology; Proceedings of the Fifth Energy Technology Conference, Washington, D.C., April, Government Institutes, Inc., R. I. Hill, Ed.

Appendix A: Coal Mines in Colorado

This list includes new mines in preparation and mines in production, as well as mines that are temporarily inactive, or permanently closed mines that still retained a 1977 license. However, mines licensed in 1976 which closed that same year will not be listed if a new 1977 license was not requested.

COLORADO

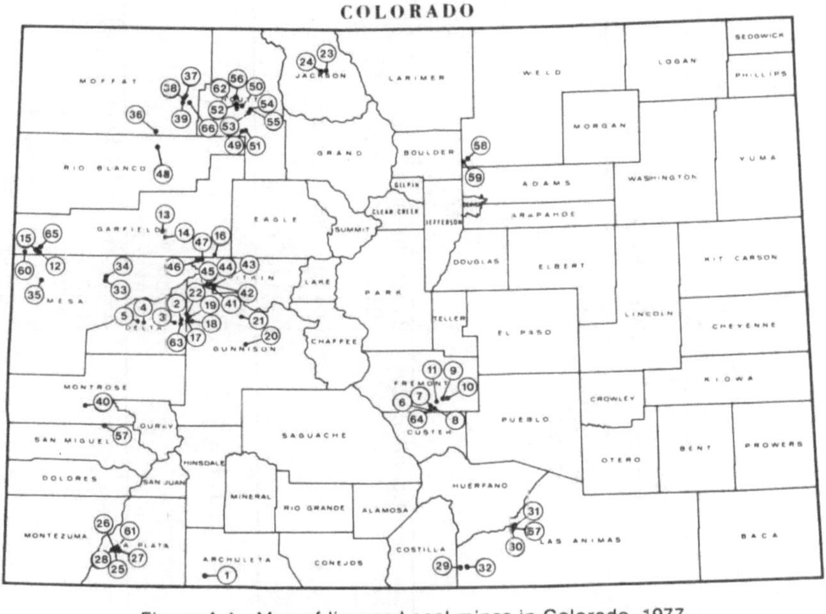

Figure A.1 Map of licensed coal mines in Colorado, 1977.
Source: Dawson and Murray, 1978, p. 27.

TABLE A.1

LOCATIONS OF COAL MINES LICENSED AS OF DECEMBER 31, 1977

Map No.	Coal Region	Mine Name	Surface (S) or Underground (U)	Location Sec.	Location Twp.	Location Rge.
1	San Juan River	Martinez	S	30	34N	4W
2	Uinta	King & Tipple	U	15	13S	91W
3	"	Orchard Valley (Converse)	U	24	13S	92W
4	"	Red Canyon #1 (Coalby #2)	U	12	13S	95W
5	"	Tomahawk Strip	S	10,15,16	13S	95W
6	Canon City	Black Diamond Strip (Old Corley, New GEC)	U/S	24	20S	70W
7	"	Cedar Canon Strip	S	35	19S	70W
8	"	Golden Quality #5	U	2	20S	70W
9	"	Hastings Strip	S	19	20S	69W
10	"	Newlin Creek	U	30,31	20S	69W
11	"	Twin Pines	U	1	20S	70W
12	Uinta	East Salt Creek	burned out	9	7S	102W
13	"	Eastside	U	24	5S	92W
14	"	NuGap #3	U	24	5S	92W
15	"	Spink Canyon	closed		7S	102W
16	"	Sunlight (Old Four Mile)	U		7S	89W
17	"	Bear	U	9,16	13S	90W
18	"	Hawks Nest East #2	U	11	13S	90W
19	"	Hawks Nest West #3	U	11	13S	90W
20	"	O.C. Mine #2	U	16	15S	86W

21	"	Peanut	U	28	13S	86W
22	"	Somerset	U	8	13S	90W
23	No. Park	Canadian Strip	S	2	8N	78W
24	"	Marr Strip #1	S	36	8N	78W
25	San Juan River	Blue Flame	U	31	35N	11W
26	"	Hay Gulch	U	36	35N	12W
27	"	King	U	32	35N	11W
28	"	Peacock	U	29	35N	11W
29	Raton Mesa	Allen	U	27	33S	68W
30	"	Healey Strip	S	21	30S	65W
31	"	Jewell Strip	S	21	30S	65W
32	"	Maxwell	U	29	33S	67W
33	Uinta	CMC (New Roadside, Old P.V., and Riverview)	U	34	10S	98W
34	"	Cameo	U	27,28,33,34	10S	98W
35	"	McGinley #1	U	5	9S	100W
36	"	Colowyo	S	2,3,4,9,10	3N	93W
37	Green River	Trapper (Craig)	S	var.	5N,6N	90W,91W
38	"	Eagle #5 (Wise Hill #5)	U	31	6N	91W
39	"	Eagle #9 (Wise Hill #9)	U	32	6N	91W
40	San Juan River	Nucla Strip	S	25,26	47N	16W
41	Uinta	Bear Creek	U	21	10S	89W
42	"	Coal Basin	U	5	10S	89W
43	"	Dutch Creek #1	U	17	10S	89W
44	"	Dutch Creek #2	U	17	10S	89W
45	"	L.S. Wood	U	8	10S	89W

TABLE A.1 (continued)

46	"	Thompson Creek #1	U		8S	89W
47	"	Thompson Creek #2	U	34	8S	89W
48	"	Rienau #2	U	29	2N	93W
49	Green River	Apex #2	U	22	4N	86W
50	"	Blazer	U	36	7N	87W
51	"	Edna Strip & Test	S	2	4N	86W
52	"	Eilts (Same as Meadows Strip #1)	S/U	23	6N	87W
53	"	Energy Strip #1	S	8	4N	86W
				32,33	5N	86W
54	"	Energy Strip #2	S	19,30	5N	86W
				25	5N	87W
55	"	Energy Strip #3	S	1,2	5N	86W
56	"	Seneca #2	S	1,2,3	5N	87W
				34,35,36	6N	87W
57	San Juan River	Elder	U	20	45N	13W
58	Denver	Eagle	U	15	1N	68W
59	"	Lincoln	U	24	1N	68W
60	Uinta	McClane Canyon		21	7S	102W
61	San Juan River	Coal Gulch (Old Victor)	U	15,16,20, 21,22	35N	10W
62	Green River	Meadows Strip #1 (Eilt's Property)	S	23,24,26	6N	87W
63	Uinta	Blue Ribbon	U	2	13S	91W
64	Canon City	GEC (Old Black Diamond)	S		19,20S	69W
65	Uinta	Munger	U	27	7S	102W
66	Green River	Wms. Fork Strip #2	S	30,31	6N	91W
67	Raton Mesa	Delagua Strip (Berwind)	S	15	31S	65W

<u>SOURCE</u>: Dawson & Murray, 1978, pp. 26-29.

Part II

Oil Shale

11 Nature of the Resource

OIL SHALE: PHYSICAL DESCRIPTION AND CHARACTERISTICS

Called the "rock that burns" by frontier Indians, Western oil shale is a misnomer since it is neither a shale nor does it contain oil. Oil shale is, instead, an organic marlstone--a sedimentary rock containing solid organic materials in a mineral matrix. It is these organic materials which are the source of crude shale oil.

About 5-10 percent by weight of the organic material in oil shale is bitumen, a tar-like substance soluble in many conventional organic solvents (Hubbard and Robinson, 1950). The remaining organic material is kerogen, a hydrocarbon molecule of high molecular weight having strong rock-binding properties.

Unfortunately, because of the chemical properties of oil shale kerogen, conventional solvent-extraction techniques can remove only a small fraction of the organic material (Ferris, 1948, pp. 12-22). When oil shale is heated, however, the organic materials decompose to form gaseous and liquid products, including crude shale oil; the nature and extent of these products are shown in Table 11.1. The heating process which converts the bitumen and kerogen to oil and gas is called pyrolysis, or retorting; the vessel in which the oil shale is heated is called a retort.

A standard laboratory procedure, known as the Fischer assay, is used to find the amount of oil contained in an oil shale sample. This technique involves heating a 100-gram sample of oil shale. This is done in an aluminum retort, without air, and according to a standard heating program (Sladek, 1974, p. 6). The Fischer assay technique permits reporting of the grade of the oil shale sample in gallons of oil per ton of shale (GPT).

Oil shale is actually a combination of several organic and mineral elements, as illustrated by Table 11.2. In addition to these elements, the oil shale deposits in the northern portion of the Piceance Creek Basin in northwestern Colorado contain scattered, but often large, deposits of dawsonite, nahcolite

TABLE 11.1

PRODUCTS FROM PYROLYSIS OF 26.7 GALLON PER TON COLORADO OIL SHALE (at 900° F)

Pyrolysis Product	Weight % of Constituent in Raw Shale	Weight % of Total Raw Shale
Oil	63	10.4
Noncondensable gas	15	2.5
Fixed-carbon residue	13	2.2
Water vapor	9	1.4
Total	100	16.5

SOURCE: Allred, V. D., 1967.

and other saline minerals. Dawsonite (dihydroxy sodium aluminum carbonate) is a potential source of aluminum. Found occasionally in thin crystalline layers, dawsonite more commonly occurs as microscopic crystals scattered through the oil shale. Nahcolite (sodium bicarbonate) is a source of soda ash, a common commercial chemical used in many products and many industrial processes, including glass production. One major use of nahcolite could be as a stack-gas scrubber in flue gas desulfurizer units for coal-fired power plants. When used dry, nahcolite is capable of removing almost all sulfur dioxide and some of the nitrogen oxides produced in coal combustion (See Coal, Part I). Nahcolite occurs in selected zones of Colorado's oil shale, appearing in aggregate clumps, or lenses, which may range from a few inches to several feet in thickness.

Hite and Dyni (1967, pp. 25-38) examined the core drillings of a 1,340 foot corehole site in the northern Piceance Basin to estimate the resources of oil shale, nahcolite, and dawnsonite. In one zone the average yield of shale oil was more than thirty-one gallons per ton, and they calculated the shale oil concentration in the area of the corehole at nearly one billion barrels per square mile. Under the same square mile, Hite and Dyni estimated there were 130 million tons of nahcolite with a potential yield of 82 million tons of soda ash, and 42 million tons of alumina found in the dawsonite-rich zones. This compares significantly to an estimated thirty million tons of alumina found in all known bauxite deposits in the United States.

The findings of Hite and Dyni suggest the recovery of minerals, along with shale oil, may be of advantage for oil shale processing and production. However, mineral recovery may pose other difficulties to the recovery of shale oil--such as

TABLE 11.2

CHEMICAL ANALYSES OF THE MINERAL AND ORGANIC FRACTIONS OF GREEN RIVER OIL SHALE

Mineral Constituents

Mineral	Chemical Formula	Weight % of Total Minerals
Dolomite	$CaMg(CO_3)_2$	32
Calcite	$CaCO_3$	16
Quartz	SiO_2	15
Illite	(silica clay)	19
Low-albite	$NaAlSi_3O_8$	10
Adularia	$KAlSi_3O_8$	6
Pyrite	FeS_2	1
Analcime	$NaAlSi_2O_6H_2O$	1
Total		100

Organic Constituents

Element	Weight % of Total Organics
Carbon	76.5
Hydrogen	10.3
Nitrogen	2.5
Sulfur	1.2
Oxygen	9.5
Total	100.0

SOURCE: Cook, C. W., 1970; Stanfield, K. E., et al., 1951.

increased water requirements in an area already scarce in water. It should also be remembered that the principal goal of oil shale extraction is to provide fuels, not minerals. Although saline mineral resources are estimated at thirty billion tons each, dawsonite and nahcolite are found only at selected sites, and are economically recoverable only in some of these.

AMERICA'S SHALE DEPOSITS

Oil shale deposits underlie much of the United States, as shown in Figure 11.1. But it is the Green River Formation deposits in Colorado, Utah, and Wyoming that compose the largest resource of contained oil shale in the world (Sladek, 1974, p. 9). Although other United States deposits in the eastern and central states underlie a much larger area, they are lean in kerogen content and occur in relatively thin beds. In fact, the estimated shale oil resource in the eastern and central United States is less than half of the Green River shales (See Table 11.3).

TABLE 11.3

SHALE OIL RESOURCES OF THE UNITED STATES* (billions of barrels)

Shale Oil Yield Range (gallons/ton)	5-10	10-25	25+
Green River Formation (Colorado, Utah, Wyoming)	4,000	2,800	1,200
Central and Eastern U.S.	2,000	1,000	n/a**
Alaskan deposits	large	200	250
Other deposits	134,000	22,250	500
TOTAL (rounded)	140,000	26,000	2,000

SOURCE: Sladek, 1974, p. 9.

*Includes oil shale in known resources, in extensions of known resources, and in undiscovered but anticipated resources.

**Estimate not available.

Table 11.4 shows estimates of oil shale resources in the Green River Formation by state and by grade (gallons per ton). Because geological knowledge is incomplete and different evaluation criteria are often used, these estimates vary widely (West, 1977, p. 522). The distinction between oil shale resources and oil shale reserves is not clear-cut mainly because of the economic uncertainty of oil shale development. The term energy resources refers to the total discovered and undiscovered deposits, while reserves refer only to those resources economically recoverable by current available technologies.

The resource base may contain several trillion barrels of oil with 600 billion barrels recoverable from deposits of 15 gallons per ton or greater in Colorado, Utah, and Wyoming, as shown in Table 11.4. Deposits of high-grade oil shale (at least 25 gallons per ton) represent 731 billion barrels of

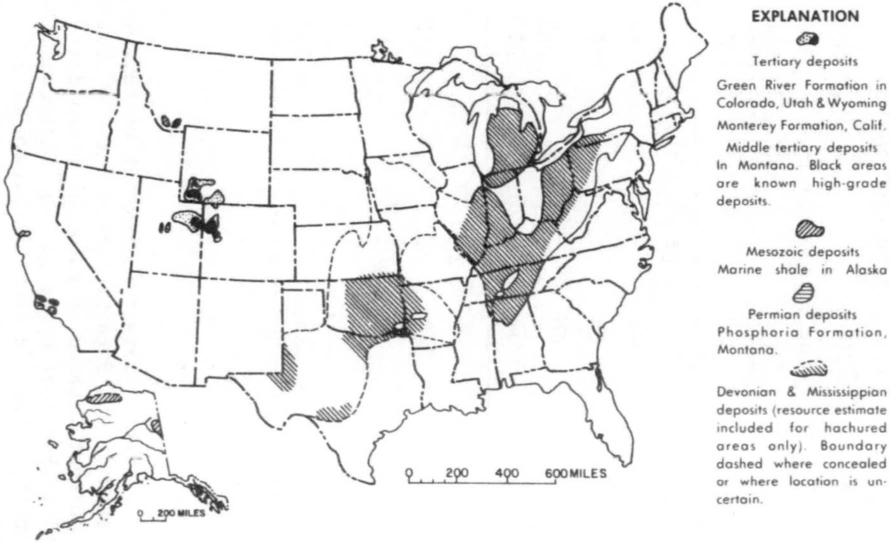

Figure 11.1 Distribution of Oil Shale resources.

Source: Environmental Protection Agency, 1978, p. 2.

TABLE 11.4

OIL SHALE RESOURCES IN BEDS AT LEAST 10 FEET THICK
(billions of barrels)

	In Place	Recoverable
	15 gal/ton shale	
Colorado	1,200	400
Utah	321	105
Wyoming	321	105
TOTAL	1,842	610
	25 gal/ton shale	
Colorado	607	202
Utah	64	21
Wyoming	60	20
TOTAL	731	243
	30 gal/ton shale	
Colorado	355	118
Utah	50	17
Wyoming	13	4
TOTAL	418	139

SOURCE: West, 1977, p. 523.

in-place deposits. Approximately 83 percent of this high-grade resource is located in the Piceance Creek Basin of Colorado, with about 8.5 percent located in both the Uinta Basin of Utah and the Green River Basin of Wyoming. The Piceance Creek Basin, along with ownership areas, is shown in Figure 11.2.

The U.S. Department of Interior reports the Federal government is the largest owner of oil shale lands, holding nearly 72 percent (about 7.2 million acres out of more than 11 million acres judged suitable for commercial development in Colorado, Wyoming, and Utah); these lands hold about 80 percent of U.S. oil shale resources (West, 1977, p. 524).

COLORADO'S OIL SHALE DEPOSITS

As the grade of oil shale increases, Colorado's percentage of both in-place and recoverable reserves also increases, as illustrated in Table 11.4. Thus, Colorado has the nation's thickest, richest, and best-defined oil shale resources. Most of these resources are located in the area of the Piceance Creek Basin of northwestern Colorado (Garfield, Mesa, and Rio Blanco counties). A list of the tracts nominated for the prototype federal oil shale leasing program in Colorado, along with estimated reserves for each tract is included in Appendix B. Two tracts have been leased to date (See Table 11.5).

TABLE 11.5

FEDERAL OIL SHALE TRACKS LEASED IN COLORADO

Tract*	Lease Date	Estimated Recoverable Resource (millions of barrels)	Acreage	High Bid) ($ millions)
C-a	1-8-74	4,070	5,090	210.4
C-b	2-12-74	723	5,090	117.8

SOURCE: Strang, M. L., 1974, p. 44.

*Both tracts are located in Rio Blanco County.

The Piceance Creek Basin has three separate oil shale zones: a rich, upper mahogany zone (called so because of its mahogany coloring); an intermediate porous zone; and a rich, lower zone. The mahogany zone contains an estimated 164 billion barrels of shale oil. The thick, rich portions of this zone, located in the heart of the basin, have a total estimated resource of 42 billion barrels, with an average of 350,000 barrels per acre, or over 2.2 million barrels per square mile. The thickness of the mahogany zone varies from less than 75 feet in the southern part of the basin to more than 225 feet in the northern part; the overburden thickness (the material covering the shale) ranges from less than 200 feet in the northwest corner to more than 1,000 feet in the center of the basin. The oil shale outcrops on the southern escarpments of Roan and Parachute Creeks, Dow (Colony) and Union, for example, have mines on the side of the cliffs entering the mahogany zone directly. This is also true at Anvil Points (Paraho) (Hutchins, 1978).

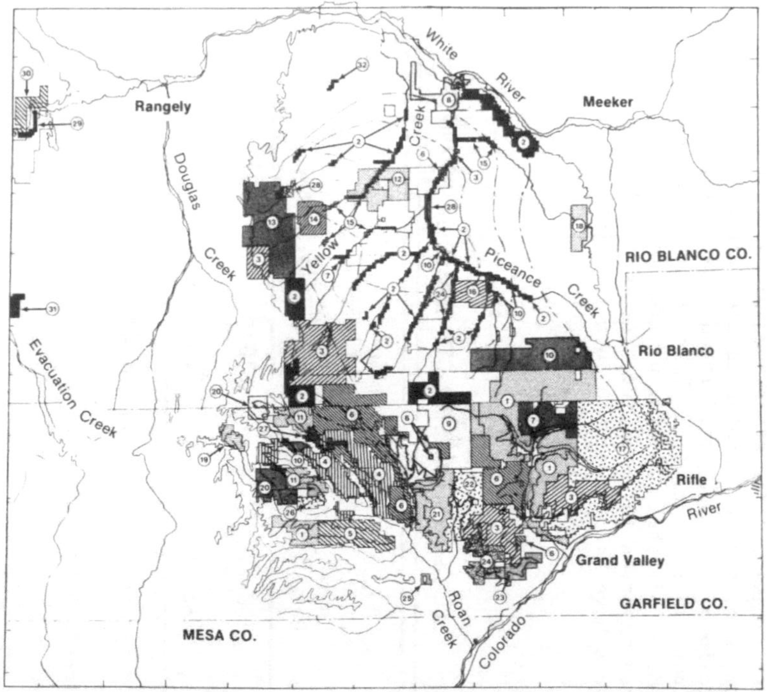

Figure 11.2 Colorado Oil Shale Areas (illustrated by ownership).

Source: Derived from Cameron Engineers Map 1978.

OWNERSHIP

1. Union Oil Company
2. Exxon, et al
3. Mobile Arco Equity, et al
4. Texaco
5. Conoco
6. Chevron Shale Oil Co.
7. Colony Development
8. Superior
9. Getty
10. Tosco, et al
11. Savage
12. Industrial Resources
13. ERTL
14. Rio Blanco Oil Shale Project (Gulf & Standard)
15. Bell, et al
16. Occidental & Ashland
17. Naval Oil Shale Reserves
18. Nelson
19. Woodruff
20. Meserve
21. Sohio
22. City Service
23. D.A. Shale
24. Shell, et al
25. Farnum
26. Hugg
27. Kerogen
28. Marathon
29. Icy Stateler
30. Engineers Oil Shale
31. Utah-Colorado Oil Corp.
32. Caldwell

The intermediate oil shale zone, buried more deeply than the mahogany zone, is less thoroughly explored. The oil shale deposits in this zone vary in thickness from 100 to 250 feet; the overburden ranges from about 300 feet in the northwest corner to 1,250 feet in the center of the basin. The rich, lower zone extends about 2,000 feet below the surface near the center of the basin and ranges in thickness from 100 to 150 feet. Overburden thickness, including the oil shale zones above, varies from 1,150 feet in the southeast corner of the basin to 1,800 feet in the center. In the northern part of the basin, where the richer zones overlap, the shale oil concentrations exceed 1 million barrels per acre of 640 million barrels per square mile (Sladek, 1974, pp. 13-14).

WATER REQUIREMENTS AND USES

Water requirements could be a major obstacle to the large-scale development of an oil shale industry in Colorado. As with many other energy production processes, shale oil processing requires large amounts of water. For example, Table 11.6 shows estimates of the water consumption for three possible commercial shale oil operations and the associated urban populations. These water requirement estimates are for shale oil production only and do not include potential multi-mineral processing.

A plant which produced nahcolite and dawsonite minerals in addition to 50,000 barrels per day (BPD) of shale oil would require from 27,000 to 38,000 acre-feet of water per year (AF/Y). This is approximately four times the amount of water which would be required for a 50,000 BPD <u>in-situ</u> operation producing only shale oil (Project Independence Oil Shale Task Force, 1974, p. 159).

Estimates of water needs vary widely, depending on what assumptions are made and who is doing the estimating. The Project Independence figures in Table 11.6 are concise, thorough and clear; however, other recent estimates may show slightly different figures. Table 11.7 shows a summary of various estimates concerning water utilization for a 50,000 BPD oil shale facility. Actual water needs can only accurately be known when a full scale oil shale development is in production, although other estimates indicate that for a 50,000 BPD plant, 3-6,000 AF/Y is required for surface processing and 2-4,000 AF/Y for <u>in-situ</u> processing (Hutchins, 1978).

Water for oil shale development could be drawn from four general sources: (1) ground water proven not part of the allocated surface water system which would take rigorous geologic proof; (2) appropriations of surface water already held by the companies; (3) appropriations of surface water held by agriculture or ranches, previously purchased as a property right and allowed by a water court to be converted to energy use; (4) water produced in the processes being (a) water of crystallization (in rock) and (b) water of pyrolysis (i.e.,

TABLE 11.6

WATER CONSUMPTION FOR SHALE OIL PRODUCTION (acre-feet per year)

Use Category	50,000 barrels per day of shale oil, underground mining	100,000 barrels per day of shale oil, surface mining	50,000 barrels per day of shale oil, BuMines *in-situ* processing
Process Requirements:			
Mining and Crushing	370- 510	730- 1,020	0
Retorting	580- 730	1,170- 1,460	0
Shale Oil Upgrading	1,460- 2,190	2,920- 4,380	1,460- 2,220
Processed Shale Disposal	2,900- 4,400	5,840- 8,750	0
Power Requirements	730- 1,020	1,460- 2,040	730- 1,820
Revegetation	0- 700	0- 700	0- 700
Sanitary Use	20- 50	30- 70	20- 40
Subtotal	6,060- 9,600	12,150-18,420	2,210- 4,780
Associated Urban:			
Domestic Use	670- 910	1,140- 1,530	720- 840
Domestic Power	70- 90	110- 150	70- 80
Subtotal	740- 1,000	1,250- 1,680	790- 920
TOTAL	6,800-10,600	13,400-20,100	3,000- 5,700
AVERAGE VALUE	8,700	16,800	4,400

SOURCE: Project Independence Oil Shale Task Force, 1974, p. 154.

TABLE 11.7

ESTIMATED WATER UTILIZATION FOR AN INTEGRATED 50,000 BPCD* OIL SHALE FACILITY (acre-feet/year)

Estimator	Mining method	Retort technology	Mining & crushing	Retort-ing	Upgrading	Spent shale disposal	Revege-tation	Misc.	Total
Tract C-a	Open pit	TOSCO II	935	2,837	1,800-2,075	3,300	550	200	9,622-9,897
Tract C-b	Under-ground	TOSCO II gas combustion	904	3,137	2,216	2,418	1,128 (After 12 yrs)	450	10,523
Tracts U-a/ U-b	Under-ground	Gas com-bustion vertical retorts	630	855	1,426	2,290	1,078	850	7,129
PARAHO									
Direct	Under-ground	Paraho	630	120	1,000	0	1,430	1,000	4,180
Indirect	Under-ground	Paraho	630	150	1,400	0	3,055	1,000	6,235
Tract C-a**	RISE in situ	In situ/ surface	200	3,100	375-390	1,354	200	NA	5,229-5,794
Tract C-b	OXY in situ	In Situ	725	2,738	0-375	300	200	200	4,638
Interior Dept. Prototype lease EIS	Surface	TOSCO II	510	730	2,190	4,375	350	70	8,225
Interior Dept. Prototype lease	Under-ground	TOSCO II/Gas combustion	370 510	580 730	1,460 2,190	2,900 4,400	0 700	20 50	5,330 8,580

TABLE 11.7 (continued)

Interior Dept. Prototype lease EIS	In situ	In situ	0 0	0 0	1,460 2,220	0 0	0 700	20 40	1,480 2,960
TRW (Gov't. study)	In situ	In situ	74 102	116 146	1,460 2,214	580 880	0 700	20 40	2,250 4,082
Superior	Under-ground	Circular grate	NA	NA	NA	NA	0	NA	4,702

SOURCE: Office of Technology Assessment, 1978.

*Barrels Per Calendar Day

**Some numbers estimated by Cameron Engineers, Inc.

created by molecular hydrogen and oxygen in retort) (Hutchins, 1978). Ground water could be used for initial water needs on leases in the northern part of the Piceance Basin. However, as these supplies are depleted, other sources would have to be found. Large-scale development, therefore, must ultimately depend on the continued availability of surface supplies. Water rights can be purchased by oil shale developers like any other commodity; however there are many legal complexities in doing so (See Vranesh, 1977, pp. 34-44).

Waste disposal and shale oil upgrading (partial refining) operations together account for about 60 percent of the water requirements for a 50,000 BPD surface retorting plant, as shown in Figure 11.3. If the water consumed by both these is reduced through new technology however, available water supplies could support a higher level of shale oil production. For example, in waste disposal process, most of the water is used in compacting the spent shale. The amount of water can be reduced by using only as much as necessary to prevent dust problems in the initial stages of compaction, followed by a gradual increase of the percentage of water used in the latter stages of disposal.

Pure _in-situ_, or in-place, retort processing should not involve this waste disposal question, since this process would create little, if any, spent shale for disposal. However, _in-situ_ processing will require modest water or steam injection for retorting in-place or upgrading of the shale oil.

Another method for reducing water requirements of oil shale processing is to transport the heavy, thick crude oil via a heated pipeline or an insulated or heated tank car to another processing site, or add a pour point depressant. This would eliminate the need for local water use in upgrading. If the pipeline is buried, it is possible that the hot oil entering the pipeline may have sufficient thermal capacity to maintain a modest temperature drop along the route. This could eliminate additional heating and added insulation. Water requirements could also be reduced by importing water from other areas or by relaxing land reclamation requirements. Of course, these environmental trade-offs have their own set of impacts which have to be balanced against the benefits of oil shale development.

In summary, the availability of enough water is an important long-term constraint on the development of a _large scale shale oil industry_. However, to place water use for energy development in perspective, shale oil production may use less water per Btu output than all other energy-producing systems except for uranium fuel processing and oil refining (Davis and Wood, 1974, p. 12).

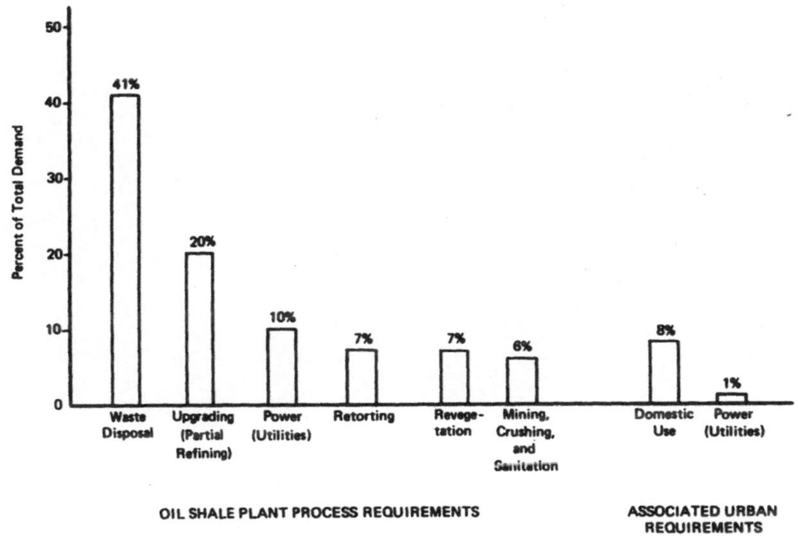

Figure 11.3 Process water requirements for a 50,000 BPD plant.

Source: Project Independence, 1974, p. 295.

12
Oil Shale Mining

The only practical method presently available for recovering oil from oil shale is heating the shale to temperatures high enough to decompose kerogen--yielding shale oil, gas, and residue. Options, however, do exist in the methods; they are:

- Open pit mining and surface retorting;
- Underground (room and pillar) mining and surface retorting;
- <u>In-situ</u> processing;
- Modified <u>in-situ</u> processing (the initial void space is provided by excavation).

(See next chapter also for details on retorting.)

The decision about which method to use depends on several technologic and economic factors: overburden depth, seam thickness, deposit size and other site-specific geologic and hydrologic characteristics, and environmental considerations.

OPEN PIT MINING

Oil shale is much harder* than coal; its deposits, or seams, are also generally much thicker than coal seams. Consequently, an oil shale surface mine is more like a limestone quarry or an open-pit copper mine than a coal strip mine. The same basic coal mining activities are used, however (Hebb and Morse, 1976, p. 14). These techniques include surface preparation, fracturing (drilling and blasting), excavation and ore removal. Presently, there is little information available about actual surface oil shale mining operations, and the following discussion is based upon a study of proposed surface operations on Tract C-a (Banks and Franciscotti, 1976).

---

*Compressive strength range is 15-25,000 psi; it varies inversely with kerogen content (Hutchins, 1978).

Selection of a specific site for an open-pit operation depends on such factors as the resource concentration, overburden thickness, subsurface water quality, potential environmental disturbance and early production of retort feedstocks (crude shale oil for refining) at minimum costs.

Water is a major consideration in mining oil shale. First, as the shale is taken from the ground, ground waters may seep into the mine. As mining continues, this water must be removed through a process called dewatering. In fact, on federal tracts C-a and C-b, dewatering requirements were estimated to be 55 acre-feet per day from a potential mine two miles wide. This inflow of ground water was greater as mining advanced toward the higher resource areas.

Surface Disturbances

One of the prime objectives of mining design is to minimize surface disturbance. In open pit mining this can be done with the single-pass open-pit mining technique. This method represents a one-time, interim land use, and allows for prompt restoration of previously mined areas as the appropriate pit size is reached. With the single-pass technique, mining moves in one direction across the bench, or flat excavation area. Waste materials and overburden are used to back-fill the mined-out areas behind the working face of the mine.

Before this technique is used, however, the pit must be developed. As this is done, the ore and waste materials are crumbled by in-pit crushers. Since the pit may range from 1,300 to 1,900 feet deep, conveyors are needed to remove the ore and waste materials from the pit. After the pit reaches operating depth with the mining complete on one face, and after adequate space is available to operate the equipment safely, tram-roads may be built to return the waste material to the already-mined areas. Therefore, land restoration can begin in one area while the mining progresses in another.

The amount of waste material is, of course, a crucial consideration in any mining operation. For oil shale surface mines, Table 12.1 shows typical waste-to-shale ratios for three possible cut-off grades of oil shale (a cut-off grade is established by stating the minimum acceptable grade, or oil shale yield in gallons of oil per ton of shale; oil shale deposits containing less than this minimum grade are considered waste material). Of course, as the cut-off grade decreases, the amount of waste material as well as the waste-to-shale ratio increases. More of the resource can be captured through surface mining, as the economic cut-off is much lower. Underground mining will be used for only higher grade ore. _Thus there is an environmental trade-off in surface mining_: disturbance of the land against utilization of the resource.

A benefit of open pit mining is that it might make recovery of lower grade deposits economical since this material would have to be removed in any case if it lies above the

TABLE 12.1

TYPICAL WASTE-TO-SHALE RATIOS FOR SURFACE MINING (1.25 Million Tons per Day)

	Oil Shale Cut-off Grade in Gallons Per Ton (GPT)								
	at least 15 GPT			at least 20 GPT			at least 25 GPT		
Mining Year	Total Tons (MM)*	Avg. Grade (GPT)	WS**	Total Tons (MM)*	Avg. Grade (GPT)	WS**	Total Tons (MM)*	Avg. Grade (GPT)	WS**
1	113	-	-	113	-	-	218	-	-
2	322	-	-	322	-	-	620	-	-
3	528	19.8	13.2	491	23.1	18.6	1022	-	-
4	692	19.1	2.9	680	23.5	8.2	1408	28.3	99.5
5	700	20.1	1.8	866	23.6	5.1	1800	26.7	30.6
6	691	20.3	1.1	1054	23.3	4.0	2187	27.4	16.6
7	764	20.6	0.9	1279	24.0	3.4	2576	28.5	13.2
8	808	20.2	0.8	1318	24.5	2.7	2775	28.8	8.7
9	808	21.1	0.8	1395	24.6	2.8	2756	28.9	5.7
10	808	21.1	0.8	1411	24.7	2.2	2752	28.2	5.4
20	875	22.0	1.0	1457	24.5	2.3	2755	28.0	5.4

SOURCE: Banks, C. E. and Franciscotti, B. C., 1976.

*Total ore and waste, in millions of tons.
**WS - waste-to-shale ratio in tons waste per ton oil shale above cutoff grade.

primary (cut-off grade) zone. In any event, as the amount of waste, or overburden, increases, so does the total investment required. And, at some point, it is more economical to begin an underground mining operation than it is to remove the overburden for an open pit mine (See also Environmental and Commercial Development chapters).

UNDERGROUND MINING

The most well-developed mining technology for oil shale recovery is the room and pillar technique, a method also used in underground coal mining (See Coal, Part I). Since oil shale seams are both thicker and harder than coal seams, the conditions for underground mining are better. These include better roof conditions for oil shale, the absence of explosive gases, reduction of dust, and less tendency of roof caving than in coal mining.*

To date, actual underground oil shale mining operations have only been done on a pilot scale. Thus, since information about commercially-viable underground operations is not yet available, this discussion relies on a recent study (Hoskins et al., 1976) for the following examination of underground mining. The study, performed under contract to the U.S. Bureau of Mines, addressed the technical and economic feasibility of mining deep, thick oil shale deposits in the central portion of the Piceance Creek Basin. Some design constraints of the study were:

- Production tonnages would be equal to or greater than 85,000 tons per day (TPD).
- Present health and safety standards must be met.
- Mine designs must be capable of handling large inflows of water.
- Underground disposal of spent shale must be considered.
- High resource recovery must be sought.

---

*Extensive additional tests are needed to determine (and forecast) long-term pillar load-carrying capabilities and stress distribution in the mine. Colony Development Operation and others are investigating a rock mechanics program (Ludlam and Nutter, 1977, p. 18). Uncertainties still exist about the costs of underground haulage of mined rock and the effectiveness of ventilation systems in oil shale mining. Many of these uncertainties will be removed, of course, through operation of a large-scale underground mine.

- Mining environment included: overburden, 1,000 ft.; oil shale, 2,000 ft.; 600 ft. thick aquifer located 1,200 ft. below surface.

The study also made correlations between certain engineering properties of Green River Oil shale and Fischer assay (gallons of oil per ton of shale) data. In the past, engineers have relied on corehole samples to determine these engineering properties. However, after many years of studying these properties along with Fischer assay results, engineers can now use the Fischer assay to determine the engineering properties of a particular shale deposit. For example, a high kerogen content indicates the shale deposit is very strong and less apt to be fractured. Properties examined in this study included: compressive and tensile strengths, elastic deformation, cohesion and creep. Further, the study data provided a basis for pillar and roof beam design for an underground mine and enabled prediction of subsidence.

Factors for an Underground System

In selecting a mining system, the U.S. Bureau of Mines study also evaluated and weighted these factors to rank the candidates:

Technical Feasibility

- Equipment development
- Preproduction time
- Stage of method development
- Productivity

Mining Cost

- Capital investment
- Labor and supervision
- Operating cost

Resource Recovery

- Mining selectivity
- Percent extraction

Reclamation

- Socio-economic
- Wildlife
- Land use
- Underground spent shale disposal

Health and Safety

Relevant conclusions drawn from the study are:

"Large scale, underground mining of the central Piceance Creek oil shale deposits is technically feasible and low in cost. Four of the six mining systems evaluated were selected as being the most promising for further evaluation. These mining systems include:

- Chamber and pillar.
- Sublevel stopping with spent shale backfill.
- Sublevel stopping with full subsidence.
- Block caving using load haul dumps (LHDs).

"Projecting from available hydrologic data, water in-flow from the three aquifer systems will not significantly impede oil shale mining in the representative mine site, although it can be expected to add to mining costs. An expected flow rate of 10,000 gallons per minute (GPM) can be adequately handled with present technology.

"The rock mechanics data on Green River oil shale in the central portion of the Piceance Creek Basin are insufficient to determine design safety factors or failure probabilities. A physical testing program is needed to provide more accurate and reliable design data."

13
Shale Oil Processing

Once oil shale is mined, it must go through several processing steps before it produces a useful petroleum feedstock. As shown in Figure 13.1, after mining, the oil shale must be crushed and retorted to produce crude shale oil; the spent shale must be disposed of; and the shale oil must be upgraded, or partially refined, but not necessarily on site. The end product of this processing is a premium synthetic crude oil, or syncrude, that can easily be transported to a refinery where it can be refined into a wide range of products such as gasoline, jet and diesel fuels and domestic and industrial heating oils.

To date, retorting is the only commercially attractive method for extracting raw shale oil from oil shale deposits. In retorting, the oil shale is heated until it breaks down into crude shale oil, its by-products and waste materials. Three general retorting techniques have been developed for use on oil shale: (1) surface retorting, in which crushed oil shale is injected into a heating vessel on the surface; (2) "pure" in-situ retorting, which calls for the creation of an in-place (underground) retort and injections of natural gas as one method to start the heating process; (3) modified in-situ retorting, which combines elements of both surface and in-situ retorting.

SURFACE RETORTING

A surface retort is a heating vessel situated aboveground. Crushed oil shale is placed in this retort and heated to a temperature between 900° F and 1500° F depending upon the particular retorting process used. The heated oil shale decomposes to produce crude shale oil, a by-product hydrocarbon gas and spent shale. Three basic surface retorting methods, classified according to the way the oil shale is heated, are being examined for possible commercialization. These are: (1) the TOSCO II process, which uses hot solids; (2) the Union, Paraho, and the Bureau of Mines Gas Combustion retorts,

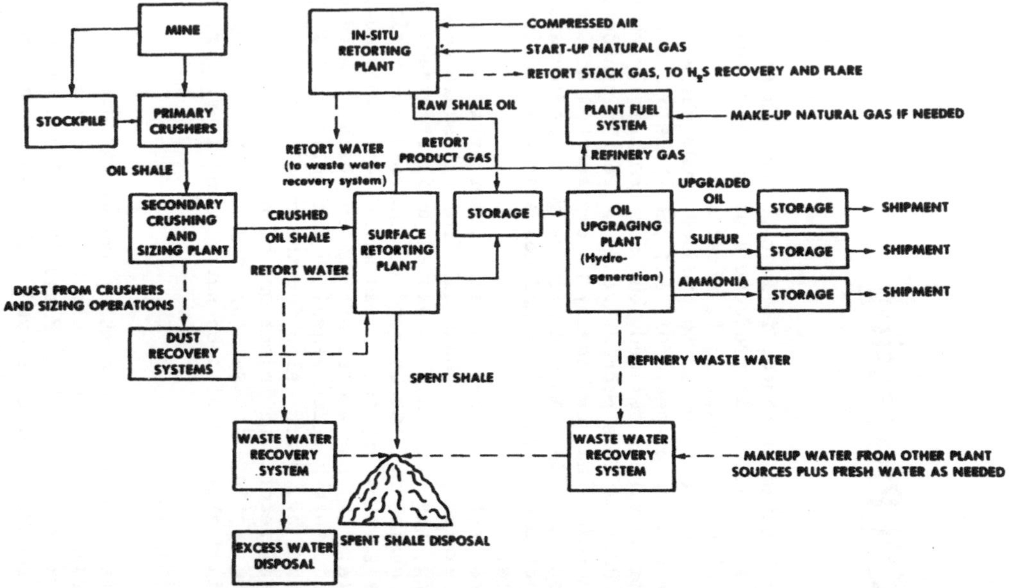

Figure 13.1 Flow diagram - Oil Shale processing.

Source: Project Independence, 1974, p. 311.

which generate an internal combustion zone within the retort; and (3) the Petrosix process, where heated gas comes from an external, fuel-fired furnace. Table 13.1 summarizes all of the processes. All the processes except TOSCO II use vertical retorts. The TOSCO II retort uses externally-heated ceramic balls in a rotating horizontal kiln.

Types of Retorts

As mentioned before, the oil shale must be crushed before it is fed into surface retorts--and the degree of crushing required varies with the type of retort. Vertical retorts generally accept a moderately large raw shale feed--that is, lumps not exceeding three inches in diameter (<3"). Hence, crushing costs are relatively lower. However, material smaller than 1/4 inch (commonly called the "fines") must be screened out. If not, these small pieces cause channeling of the gaseous heat carrier--the heat gets diverted into a single stream rather than being diffused throughout the retort, resulting in low oil recovery. These "fines," which make up about 4 percent of the crushed shale, can be discarded, handled in a special auxiliary unit using the TOSCO II process, or compressed into pellets or briquets and then put through the vertical retort. Briquetting, however, is a costly procedure. For instance, a 50,000 BPD shale oil complex proposed by the Bureau of Mines would include a briquetting plant to compress the fines for direct feeding to the retort. But, while the briquetting plant would allow use of all of the crushed oil shale, it also would add nearly $1.7 million to the capital cost of this proposed plant (Katell and Wellman, 1974). This is roughly equivalent to $2.5-$3.0 million in 1978 dollars (Hutchins, 1978).

Smaller "fines" can be handled by the Union "A" vertical retort which has a unique design feature--a "rock pump" that pushes shale up through the retort. This rock pump handles fines down to 1/8 inch and it reduces channelization. However, it is more susceptible to mechanical breakdown than the vertical retorts (Carver, 1964) which use a "gravity feed" where raw shale is introduced at the top of the retort and spent (processed) shale is removed from the bottom.

The rotating horizontal kiln used in the TOSCO II process will accept oil shale feeds of less than one half inch in size, without having to remove the fines. Additionally, the retort off-gases are not diluted with combustion products and thus provide a higher quality fuel gas than the other retorts. However, the spent shale generated by the TOSCO II process consists of very fine particles requiring greater care for dust control and disposal.

The amount of oil recovered in surface retorting methods depends upon the process used, as illustrated in Table 13.2. In general, however, recovery efficiency exceeds 75 percent of the hydrocarbons available.

TABLE 13.1

OIL SHALE RETORTING SCHEMES

Process Developer	Process	Demonstrated Throughout, T/D	Announced Consideration of Its Use by U.S. Oil Shale Developers
Petrobras and Cameron Engineers	Petrosix	2,000	None
Development Engineering, Inc.	Paraho	280	White River Shale Project Rio Blanco Oil Shale Project
Lurgi	L-R	15	None
Occidental Petroleum Corporation	Oxy Modified _In-Situ_	(500 B/D)	C-b Shale Oil Project
Superior Oil Company	Superior Multi-Mineral Process	100	Superior Oil Company
Tosco Corporation	TOSCO II	1,000	Rio Blanco Oil Shale Project Colony Development Operation C-b Shale Oil Project
Union Oil Company of California	Union A Union B	1,200 5	Union Oil Company of California
U.S.S.R.	Galoter Kiviter	550 275	None None

SOURCE: Ludlam and Nutter, 1977, p. 19.

TABLE 13.2

APPROXIMATE SURFACE-RETORT OIL RECOVERY EFFICIENCIES
(Percentages) (100 percent is Fischer assay shale oil content)

TOSCO II	Bureau of Mines	Paraho	Union "A"	Union SGR*
99	82-87	Over 90	84	100

*steam-gas recycle modification.

SOURCE: Sladek, T. A., 1975, pp. 6-12.

In-Situ Retorting

Oil shale can also be retorted *in-situ*, or in-place. *In-situ* retorting involves two critical operations: (1) the establishment of permeability in the oil shale formation, that is, the creation of paths to permit movement of gas and/or liquid; and (2) the introduction and control of heat in the formation.

Permeability

Some oil shale deposits are naturally permeable--they have natural fissures. For instance, portions of the oil shale beds in the Piceance Creek Basin contain deposits of water-soluble saline minerals, such as dawsonite and nahcolite; ground water circulating through the natural fissures in the beds has leached out the soluble minerals and left voids in the deposit. These permeable areas are ideal for *in-situ* retorting, once the saline ground water is pumped from the formation, providing that the permeability is uniform, the structure is sufficiently dense and horizontally oriented.

Unfortunately, most oil shale deposits are not sufficiently permeable for large-scale *in-situ* retorting. Consequently, permeability must be increased by enlarging existing fractures or creating new ones. In theory, a pattern of boreholes could be drilled into the oil shale deposit; the formation could then be fractured hydraulically or with liquid chemical explosives. Ideally, the fractured oil shale would be broken into small pieces, about the size of the crushed oil shale fed into a surface retort.

Heating

In the *in-situ* process, after creating permeability, heat is introduced into the formation. Proposed heat sources include: (1) combustion of the oil shale itself, initiated by burning natural gas or other fuels and sustained by the injection of air; (2) hot natural gas or methane; (3) hot carbon dioxide; (4) superheated steam; (5) hot solvents; or (6) combinations of two or more of these alternatives (Strang, 1974, p. 52). The combustion front of hot liquids or gases is then swept horizontally through the formation to drive the retorted shale oil out of one or more of the boreholes.

"Pure" or true _in-situ_ technology involves no mining (modified _in-situ_ methods do). It is still under experimental development as an oil-shale retorting method and has only been marginally successful to date. However, _in-situ_ processing offers many advantages over mining and surface retorting: (1) it takes one-third as many people to operate the process; (2) it takes only one-half to two-thirds the amount of water; (3) it leaves almost no spent shale for disposal; and (4) it costs less to operate (Project Independence Oil Shale Task Force, 1974, pp. 274-275).

Additionally, _in-situ_ processing may allow increased resource recovery from low-grade oil shale deposits. Most surface retorts require 25-30 gallons per ton of oil shale to be economically successful. But nearly 1.2 trillion barrels of the total 1.8 trillion barrel shale oil resource consist of low-grade deposits with seams more than 10 feet thick having an average yield of only 15 to 20 gallons per ton (Project Independence Oil Shale Task Force, 1974, p. 274). These low-grade deposits may never be used with conventional mining techniques but might be used once _in-situ_ techniques are developed.

MODIFIED _IN-SITU_ RETORTING

Modified _in-situ_ processing is actually mining-assisted _in-situ_ retorting. A vertical, underground retort is prepared by mining enough oil shale from the bottom of the shale deposit to create a room. Blastholes are then drilled upwards through the ceiling of the room into the formation. Explosives are detonated in the blastholes to gain the desired permeability. The broken rock fills the space, creating a chimney of rubble. The mine tunnel is sealed; gas supply and exhaust lines are installed; and the top of the shale rubble is ignited through a borehole drilled from the surface.

During retorting, shale oil flows to a sump, a reservoir at the bottom of the underground retort, from which it is pumped to the surface. The oil shale mined from the cavity, about 15 to 25 percent of the retort chamber, has to be discarded as waste or retorted using a surface retort. Since this spent shale must be disposed, the modified _in-situ_ process sacrifices some of the cost and environmental advantages associated with "pure" _in-situ_ retorting.

Currently, a modified _in-situ_ operation is being conducted in the Piceance Basin by Occidental Oil Shale, Inc., a subsidiary of Occidental Petroleum (Ridley, 1974). Occidental has fired several pilot retorts. For example, Occidental began a series of 500 BPD modified in-situ demonstration operations in December 1975. Shale oil production in March 1976 averaged 300 BPD; production in April 1976 averaged 575 BPD. A total of 27,500 barrels of shale oil had been recovered by the end of the demonstration series on June 30, 1976. A fifth retort is presently being tested. Occidental appears satisfied with the technical and economic feasibility of the project and is proceeding toward a commercial-level demonstration.

Table 13.3 shows the number of retorts required to extract the oil contained in shale if a production rate of 50,000 BPD is to be realized from Occidental's modified in-situ process.

TABLE 13.3

COMMERCIAL PRODUCTION OF SHALE OIL (based on Occidental Modified In-Situ Process)

Basis:			
Average Fischer Assay of Shale (gpt)	15	20	25
Specific Gravity of Shale*	2,425	2,330	2,245
Quantity of Shale Mines (ton/day)	55,566	42,425	33,939
Oil Production Per Retort (BPD)	218.3	279.3	335.6
No. of Retorts Required	229	179	149
Area Affected by Retorting (yd^2)	500,000	390,000	325,000

SOURCE: Shih, N., 1978.

Basis:

Retort Size - 120 ft x 120 ft x 250 ft
Mined Void Volume = 120 ft x 120 ft x 120 ft x 50 ft
Wall Thickness = 20 ft
Retort Burn Rate = 0.54 in/hr
Total Commercial Production = 50,000 BPD

*Average specific gravity of oil shale samples from Anvil Points area. Synthetic Fuels Data Handbook, Cameron Engineers, Inc., 1975, p. 10.

CURRENT SITES IN OPERATION AND ASSESSMENT OF TRENDS

Several oil shale developments are currently in operation in the Green River Formation, and more are scheduled to begin operation in the years to come. Figure 13.2 pictures locations of most of the present and projected operations. The four federal tracts which have been proposed for development by

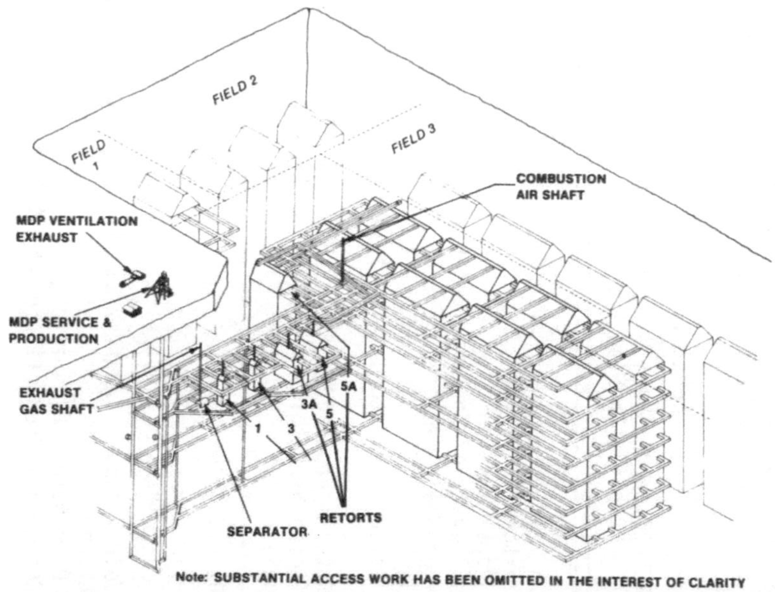

Figure 13.2 Commercial development modified in-situ side entry Tract C-a
Source: Rio Blanco Oil Shale Company, 1978

in-situ retorting are not shown, nor is the Laramie Energy Research Center (LERC) of Department of Energy (DOE). LERC personnel are managing several of the oil shale projects and are also involved in simulating in-situ oil shale processing in two above-ground retorts. The Lawrence Livermore Laboratory of DOE which is conducting retorting simulations in two pilot above-ground retorts is not pictured either. Table 13.4 summarizes key project data. For each development project, an approximate time frame is presented. For example, Rio Blanco is contemplating a combined in-situ and above-ground retorting complex, but it is probable that only the in-situ retorts will be available before 1981. Some of the activities on these sites were described briefly during the previous discussion of retorting technologies. Capsule summaries of each of the major projects are included below (Sladek, T. A., 1978).

Federal Tracts W-a and W-b

These tracts were nominated as part of the oil shale leasing program which gave rise to the four federal tracts in the Piceance and Uintah basins. The government regards the tracts as being suitable for development by in-situ retorting. No acceptable bids were received from industry.

Federal Tracts U-a and U-b

These tracts were awarded to two consortia of Phillips Petroleum--Sunoco Energy Company, and Standard Oil Company of Ohio--during the federal leasing program. The consortia proposed to develop both sites as a single unit by underground mining and above-ground retorting, using some combination of Paraho and TOSCO II retorts. The developers halted development in late 1976, though environmental monitoring continues. Although a time schedule is unavailable, the first retorting operation is not likely to begin before the mid-1980s.

Federal Tract C-a

During the federal leasing program this tract was leased to the Rio Blanco Oil Shale Project, a joint venture of Gulf Oil Company and Standard Oil Company of Indiana. Original plans called for development by mining and above-ground retorting, but these plans were abandoned because a permit could not be obtained for off-site spent shale disposal. Present plans call for development of a combination of modified in-situ retorting and above-ground processing. Commercial operation is expected by 1987, but pilot in-situ retorts should be in operation well before that time. Figure 13.2 is a conceptual illustration of a commercial development modified in-situ process.

Plate 13.1 Shaft sinking is underway on the 5,100-acre Federal Prototype Tract C-a in Rio Blanco County, Colorado, July, 1978. (Courtesy: Rio Blanco Oil Shale Corporation)

Plate 13.2 34' Dia. service shaft, April, 1978. C-b shale oil venture. (Courtesy Occidental Oil Corp.)

TABLE 13.4

SUMMARY OF RETORT DEVELOPMENT PROJECTS

Developer	Location	Time Frame	Type of Activity
White River Shale Project	Utah (Federal Tract U-A & U-B)	Mid-1980's	Aboveground commercial processing plant
Rio Blanco Oil Shale Project	Colo. (Federal Tract C-A)	1980	Possible pilot in-situ retort after 1980; commercial by 1987
Ashland and Oxy Consortium	Colo. (Federal Tract C-B)	1980	Pilot in-situ retorts
Paraho Development Corp.	Colo. (Anvil Points)	Present	Aboveground pilot retort
Colony Development Operation	Colo.	Mid-1980s	Aboveground commercial processing plant
Occidental and DOE	Colo.	Present	Pilot in-situ retorts
Union Oil Co. of Calif.	Colo.	Early 1980s	Aboveground commercial modular retort
TOSCO Corp.	Utah	Early 1980s	Combination in-situ and above-ground retorts
Geo Kinetics	Utah	Present	Pilot in-situ retorts
Superior Oil Co.	Colo.	Early 1980s	Aboveground commercial modular retort
Equity Oil Co.	Colo.	Present	Pilot in-situ retorts
Talley-Frac	Wyo.	Present	Pilot in-situ retorts
DOE	Utah	Early 1980s	Pilot in-situ retorts

SOURCE: Sladek, 1978.

Federal Tract C-b

This tract was originally leased to Colony Development, a consortium of The Oil Shale Corporation (TOSCO), Ashland Oil, Atlantic Richfield Company (ARCO), and Shell Oil. TOSCO, ARCO, and Shell have withdrawn from the program, but have been replaced by Occidental Oil Shale. The site will be developed by modified in-situ processing. Shaft sinking began in the fall of 1977, and pilot retorts should be operating in the early 1980s.

Anvil Points Oil Shale Research Center

Development work for the U.S. Bureau of Mines (USBM) Gas Combustion retort in the 1950s and 1960s was located at Anvil Points. It is presently under lease to the Paraho Development Corporation and is being used to develop the Paraho retort. Paraho has currently completed a production run to generate large amounts of shale oil for testing by the U.S. Navy. Figure 13.3 illustrates the direct mode retorting scheme. The future of the development work is uncertain. A phase-down of the Anvil Points operation was announced in September 1978 (Ritz, 1978, p. 3). This will continue unless federal legislation is enacted to further fund their operations.

Colony Property

The Colony Development Operation had planned to develop a 47,000 barrel per day commercial oil shale complex on private lands in the Piceance Basin. Underground room and pillar mining was to have been used and the shale would have been retorted above-ground in TOSCO II retorts. A final environmental impact statement was prepared for the project, but activities were stopped in 1974 because of economic uncertainties. Recent announcement of the federal oil entitlements program, and the possibility of a substantial synfuels subsidy appear to have encouraged Colony to reopen the project and to proceed to commercial operation. If this occurs, several years would be required for site development and plant construction.

D.A. Shale Property (Logan Wash)

This land is leased by Occidental Oil Shale and has been the site of in-situ retort development since the early 1970s. Occidental is presently co-funding a retort test program with DOE. Five modified in-situ retorts have been prepared and tested to date. Retort Number 6 has been prepared and will be burned during 1978, after arrangements are completed for handling retort water and groundwater which enters the retort after the burn. Retort Number 7 is scheduled for burn during 1979.

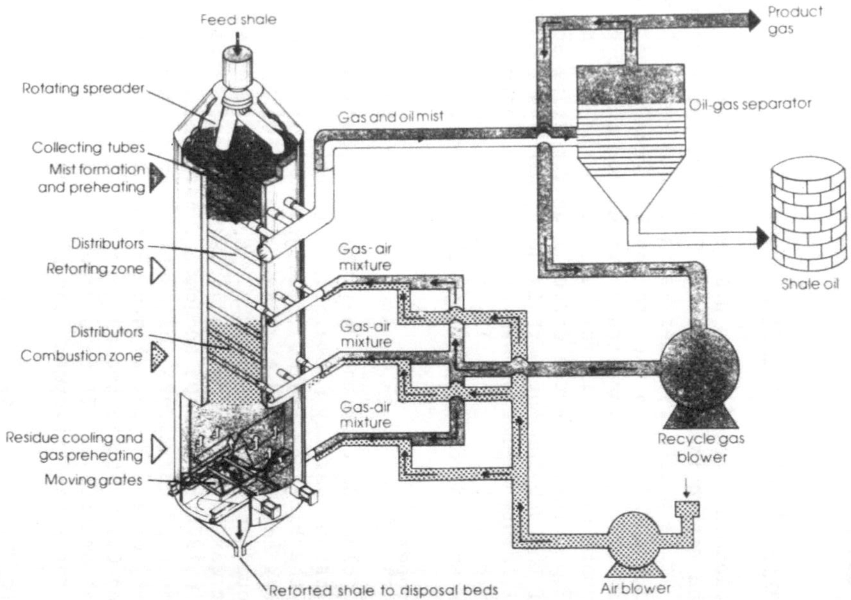

Figure 13.3 Paraho Retorting, Direct Mode
Source: U.S. Environmental Protection Agency, 1977, p. 7.

Plate 13.3 Paraho Semi-works Retort - Lump-sized oil shale is fed by a conveyor system into the top of the retort where it is heated to recover about 200 barrels a day of crude shale oil and more than two million cubic feet of usable low Btu product gas. The Paraho retort has only two moving parts, a raw shale feeder and a shale discharge mechanism. (Courtesy: Paraho Development Corporation)

Union Oil Property

This land was the site of development work on the Union Oil "A" retort during the 1950s. It is the proposed site for a modular demonstration program based on the Union "B" retort. Union is currently attempting to obtain the necessary permits for the demonstration program. Once the permits are obtained, a three-year period would be required for site development and plant construction.

Western Oil Shale Company (WESTCO)

WESTCO planned to develop state-leased tracts in the Uinta Basin by modified in-situ processing. Feasibility studies and fracture tests have been completed, but the project is currently inactive with both industry and the Federal government for lack of funds.

Sand Wash Unit (TOSCO Corporation)

TOSCO has leased a unitized cluster of state-owned lands in the Uinta Basin and plans to develop the site by a combination of modified in-situ retorting and above-ground processing in the TOSCO II retort. The target is a 75,000 barrel per day commercial complex. Environmental and exploration work is underway, but the time schedule for the project is uncertain.

Talley-Frac

Talley-Frac is conducting tests of a pure in-situ technique in the Green River Basin shales in Wyoming. The tests are co-funded by DOE. Initial fracturing of the retort volume has been completed and rubbling, leading to a burn test, is scheduled for the near future.

Geokinetics

Geokinetics has been conducting field tests on the southern fringe of the Uinta Basin since 1973. Seventeen retorts have been prepared to date, and ten of these have been burned. The development concept involves fracturing a thin bed of shale with chemical explosives and retorting the rubbled bed in a horizontal burn pattern. Work is continuing under a cost-shared contract with DOE.

Superior Oil Company

Superior plans to develop its privately-held lands in the Piceance Basin by underground mining and above-ground retorting in the Superior retort. In addition to shale oil, mineral values will also be extracted from the co-occurring nahcolite and dawsonite reserves. Work is being delayed by a land-exchange proposal between Superior and the Bureau of Land Management. If the exchange is approved and if all other factors are favorable, Superior will proceed with the demonstration project and should have a large modular retort onstream by the early 1980s.

Equity Oil In-Situ

Equity is developing a "true" in-situ retorting process on a small site in the Piceance Basin. Funds are provided under a cost-sharing agreement with DOE. Equity expected to have the initial retorting operation underway by September 1978, and to complete field testing by the end of 1980. If the Equity experiments are successful, it is likely that the field testing program will be extended and even expanded.

DOE Modified In-Situ Experiments

DOE had originally planned to conduct modified in-situ field tests in Wyoming shales but could not obtain an essential permit for road improvements from the Bureau of Land Management. Consequently, the proposed site for the tests was relocated to privately-held lands near Bonanza, Utah. The test will involve underground mining, explosive rubbling, and retorting in a horizontal forward-combustion pattern. A request for proposals for project contractors is currently being drafted and should be released within a few months. A coring program is underway at the site. The initial retort burn is planned to occur within about two years (i.e., by the end of 1980).

Additional Possibilities

In 1977, then Senator Floyd Haskell of Colorado introduced a bill (S. 419) in the U.S. Senate to authorize a program to test the commercial, environmental, and social viability of recovery technologies for domestic oil shales. The bill called for a federally-operated program to design, operate and evaluate three oil shale retorting processes, at least one of which must be in-situ. If reintroduced and enacted in its present form the legislation would allocate several hundred million dollars to the largest single oil shale venture ever conducted anywhere in the world.

The legislation was criticized on several grounds, and, if reintroduced its fate is still uncertain. However, the bill passed the Senate on June 12, 1978, by a large margin and was referred to the House where it already has gained support. If the legislation passes Congress, and if the President signs it into law, the federal program is likely to have profound impacts on development of Green River shale and upon the activities of oil shale developers.

SCALE OF OPERATIONS

All of the oil shale technologies that have been discussed, from mining to retorting, are still basically in the research, pilot, or limited production stage. For a commercial-size facility, any of these technologies must be scaled-up considerably. In fact, when considering the commercialization of oil shale, regardless of the technology used, all commercial-scale oil shale mining operations will have two things in common: they will be large, and they will be expensive.

In 1965, Tell Ertl described a commercial oil shale mine as "prodigious," because the word indicated a larger magnitude than "giant." For example, a plant yielding 100,000 barrels of crude shale oil per day would require mining 140,000 tons of 30 gallon-per-ton oil shale per day. As a comparison, consider that the largest existing underground mine in the United States, the San Miguel copper mine in Arizona, yields about 50,000 tons of ore per day. The largest open-pit mine in the United States, the Bingham Canyon pit in Utah, produces about 110,000 tons of ore per day (Sladek, 1975, p. 1). Consequently, while current technology appears capable of supporting the large-scale mining operations required by a shale oil industry, particular care is required in designing and operating the mine to minimize unit costs.

It is also important when considering commercialization to distinguish a "commercial-size" shale oil plant from a "commercial-size" oil shale retort. For example, Table 13.5 describes two conceptually different commercial-size plants: one designed in 1971 by the U.S. Bureau of Mines (USBM) as the basis for an economic analysis; the other designed in 1974 by Colony Development, which planned to build a complex on private lands (Colony's plans have been indefinitely delayed because of economic uncertainties). Notice that each complex consists of several retorts, with each retort processing from 8,000 to 13,000 tons of oil shale per day and producing between 5,000 and 9,000 barrels of crude shale oil per day. Each retort is "commercial-size," since each represents an independent production module that can be combined with other, similar modules to obtain the desired overall plant capacity.

Both the USBM gas combustion and TOSCO II "commercial-size" retorts represent a considerable extrapolation of existing technology, since they have rated capacities which are nearly ten times larger than any retort tested previously. The largest gas combustion retort tested by 1975 had a nominal capacity of 150 tons per day; the largest TOSCO II retort had a nominal capacity of 1,000 tons of shale per day (Colorado School of Mines Research Institute [CSMRI], 1975, p. V. 5). The largest surface retort of any type tested under actual field conditions was a Union "A" retort, which achieved a through-put of 1,200 tons of shale per day.

The next step in proving the suitability of current technology is to construct and operate a commercial-size retort.

TABLE 13.5

DESIGN PARAMETERS IDENTIFIED IN TWO SHALE OIL COMPLEXES

Parameter	USBM Complex	Colony Complex
1. Output	100,000 BPD (syncrude; additional refinery capacity can yield a full spectrum of refined hydrocarbons).	50,000 BPD (47,000 BPD of low sulfur fuel oil, 4,330 BPD of liquid petroleum gas, 135 TPD ammonia, 800 TPD high-ash petroleum coke, 173 TPD of sulfur.)
2. System	Three separate underground mines, each supplying shale to its own crusher. Each crusher feeds shale to a retorting train consisting of 4 Gas Combustion retorts. Oil and gas streams are combined and fed to a single upgrading facility.	Single room and pillar underground mine supplies shale to 6 TOSCO II retorting systems. Each retort includes fractionation for separation of pyrolysis vapor and into non-condensable gas, water, gas oil and heavy bottom oil. Each fraction is sent to a different process unit in the refinery.
3. Feedstocks (crushed shale)	155,000 TPD	66,000 TPD (35 gal/ton shale)
4. Spent Shale	121,000 TPD	n.a.
5. Plant Size	Commercial*	Commercial*

SOURCE: CSMRI, 1975.

*Retorts proposed for each plant represent large extrapolations of existing technology. Largest Gas Combustion retort tested to date has nominal capacity of 150 TPD; largest TOSCO II retort is 1,000 TPD.

Paraho Oil Shale Demonstration, Inc. has proposed constructing a commercial-size retort at the Anvil Points Research Center on the U.S. Naval Oil Shale Reserve. The projected daily output would be 7,000 barrels of shale oil per day.

As noted, in terms of commercialization, in-situ processing offers several advantages that are not available to mining and surface retorting operations. Unfortunately, in-situ technology is still in the experimental stage and additional research is needed to commercialize the process.

PRODUCT RECOVERY

Crude shale oil, after recovery from the oil shale deposits, must go through one more step--upgrading, or partial refining--to convert the shale oil into premium petroleum feedstock. Although this technology has not yet been tried on a commercial scale, experts feel it should not pose any insurmountable problems. The upgrading of crude shale oil is basically a modification of the upgrading process for other crudes. However, treating crude shale presents additional dimensions when considering the high nitrogen to carbon ratios of the product. Removing sulfur found in the shale oil is within the state of the art, though the presence of arsenic will undoubtedly foul the surfaces of the catalysts used in refining the oil. Nonetheless, experts believe they can solve these technical problems. In fact, some experts point out that the refining process for crude shale oil simply requires more heat and pressure than the process for other crude oils.

Shale Oil Properties

The crude shale oil produced from a surface retort differs from conventional petroleum crude oil in several ways. In general, crude shale oils are dense, viscous liquids, with a high pour point, a moderate sulfur content, and a high nitrogen content (The pour point is the lowest temperature at which the oil will flow; viscosity denotes the resistance to fluid movement). The shale oils produced by different surface retorts vary somewhat according to the process used; and the shale oil recovered from an in-situ retort is considerably different from any surface-retorted shale oil, having lower pour points, viscosities and nitrogen content. Table 13.6 compares the properties of shale oil produced by different retort processes with two types of conventional petroleum--a heavy, high-sulfur California crude and a high-quality Utah crude.

The Upgrading Process

The high pour point and high viscosity of most shale oils may result in storage and transportation problems. If the crude shale oil is not upgraded--by adding a pour point depressant, or partially refining--storage tanks or pipelines

TABLE 13.6

PROPERTIES OF SHALE OILS AND PETROLEUM CRUDE OILS

Retort Process	Gas Combustion Retort	Union "A" Retort	Union SGR* Retort	Heavy Sour** Crude Santa Maria Valley California	High Quality Crude Aneth Field Utah
Pour Point (^{o}F)	85	75	70	20	40
Viscosity (SUS)	310 at 100°F	180 at 100°F	Not Avail-able	215 at 122°F	37 at 100°F
Sulfur (weight %)	0.69	0.71	0.7	5.0	0.14
Nitrogen (weight %)	2.13	1.89	1.8	0.6	Not Available

<u>SOURCE</u>: Sladek, T. A., 1974, p. 6.

*Steam-gas recycle

**High sulfur content

will require heating to allow the flow of crude shale oil. Additionally, the high nitrogen and sulfur content of the shale oil adversely affects the refining of petroleum end products. Nitrogen may "poison" catalysts used in conventional refineries; a low sulfur content feedstock sells at a premium over a high sulfur or "sour" feedstock; and on-site upgrading or partial refining of the crude shale oil is highly desirable.

A shale oil upgrading facility would generally involve three processes: distillation, coking, and hydrocracking. In distillation, the crude shale oil is fed to a distillation column, which separates the oil into two almost-equal fractions: "light ends" or hydrocarbon vapors and "bottoms" or heavy, viscous hydrocarbon liquids. The light ends are sent to hydrocracking units which add hydrogen and break the long-chain hydrocarbon molecules into smaller, lighter molecules with lower boiling points (for example, methane, propane, butane). The bottoms are sent to delayed-coking drums where they are heated to produce more light ends and a spongy, nearly pure carbon substance called petroleum coke. Nitrogen and sulfur are removed from the gas streams at various stages, with free sulfur and ammonia recovered as by-products of the gas cleaning operations. Nitrates and sulfates, also useful as fertilizer materials, are recovered. The end product is a synthetic crude oil--syncrude--that is substantially free of sulfur, has a reduced nitrogen content, and has significantly improved flow characteristics. It is readily transportable and is considered a premium feedstock.

One proposed upgrading facility, which has an input of approximately 55,000 barrels of crude shale oil per day, would produce 47,000 barrels of low-sulfur syncrude per day; 4,330 barrels of liquified petroleum gas (LPG) per day; 135 tons of ammonia per day; 173 long tons (see Glossary) of sulfur per day; and 800 tons of high-ash petroleum coke per day (CSMRI, 1975, p. V. 4).

SHALE OIL REFINING

The differences between crude shale oil and petroleum crudes are the high nitrogen content and the unsaturated and metallic contaminants in the oil derived from shale. These lead to unusual refining problems. The contaminants prevent shale oil from being mixed with crude oils for processing in most existing refineries. However, advanced commercial state-of-the-art refining technology appears capable of refining shale oil alone or blended with petroleum crudes. An initial hydrotreating step will remove most shale oil contaminants and enable the use of conventional hydrocracking or fluid catalytic cracking. This additional step, however, will increase refining costs to about $6.50 per barrel (Sullivan, 1978, p. 1).

14
Commercial Development Consideration

LEASING OF OIL SHALE LANDS

Lands containing oil shale deposits can be leased from three sources: private land owners, State, or the Federal government. As described in the Nature of the Resource chapter, the Federal government is by far the largest owner of oil shale lands, with 72 percent of the shale acreage. Most of the remaining 28 percent of oil shale lands are owned by 18 companies in Colorado and 40 in Utah (Hutchins, 1978).

Public oil shale lands are concentrated near the geographic and deposit center of Colorado's Piceance Basin. The center of the basin holds the thickest portion of the Parachute Creek member, the best shale-bearing portion of the Tertiary Green River Formation, illustrating the Federal government's dominant position in influencing oil shale development. The Parachute Creek member ranges up to more than 1,500 feet thick under public land. By contrast, private holdings lie near the southern margin of the basin, where this member is only about 200 feet thick (West, 1978, p. 525). As a consequence, production from federal leases will form the basis for any large-scale oil shale industry development.

In 1968, the U.S. Department of Interior (DOI) began the Prototype Oil Shale Leasing Program, and an oil shale leasing study was begun in October 1969. The leasing study required the writing of an environmental impact statement (EIS) on potential oil shale development and the final draft was released in August 1973.

Beginning with competitive bonus bid sales in January 1974, the DOI offered leases on six selected tracts in Colorado, Utah, and Wyoming; during the following six months, the DOI leased four of these tracts, two each in Colorado (Tracts C-a and C-b) and Utah (U-a and U-b). Appendix C gives current information on the status of these projects. Neither of two tracts in Wyoming received acceptable bids (Environmental Protection Agency, 1978, p. A-2).

The Prototype Oil Shale Leasing Program was designed by DOI with these goals in mind:

- To provide a new source of energy to the nation by stimulating the development of commercial oil shale technology by private industry.
- To insure the environmental integrity of the affected areas and at the same time to develop a full range of environmental safeguards and restoration techniques that will be incorporated into the planning of a mature oil shale industry, should one develop.
- To permit an equitable return to all parties in the development of this public resource.
- To develop management expertise in the leasing and supervision of oil shale development in order to provide the basis for future administration procedures.

The Department of Interior must withhold further leasing of public oil shale lands until the environmental effects of the existing prototype leases are better known. A new EIS will be completed before any further leasing takes place (Project Independence Oil Shale Task Force, 1974, Appendix A).

The U.S. Geological Survey is involved in reviewing oil shale development. It maintains an Area Oil Shale Office within its Conservation Division, Central Region; this office has the prime responsibility of carrying out the goals of the prototype program. The organization structure for the office recognized that maximum participation by other agencies and by the public was one of the Secretary of Interior's desires for the program. Attaining this participation is a tremendous task because of the diffusion of expertise throughout many agencies. The Area Oil Shale Office is assisted in this task by the Oil Shale Environmental Advisory Panel created to advise Department of Interior officials about the environmental aspects of the prototype program (Cameron Engineers, 1978a, p. 24). In fact, the Area Oil Shale Supervisor (within the U.S. Geological Survey) must approve all plans of the federal lessees, and requires each lessee to submit a Detailed Development Plan (discussed in the Environmental Planning chapter).

The companies involved in oil shale development advocate a faster leasing of lands in the heart of oil shale country. They also recommend that tract sizes should be enlarged or, at least, be based on reserves and not area alone (West, 1977, p. 525). Leasing of an additional 35,000 to 135,000 acres of federal oil shale lands may be necessary by 1985, depending upon the policies of the government as to extent of industry development. In addition, off-site land will have to be made available for utility corridors for transmitting power, natural gas, water, and shale oil products as well as for construction of roads and plant facilities. Land will also be needed for processed-shale disposal areas (Project Independence Oil Shale Task Force, 1974, p. 111).

Besides the amount of acreage to be leased, methods for additional leasing must be determined. There are advantages and disadvantages to the interested parties for each possible leasing method (competitive vs. non-competitive, sealed bid vs. oral bid, royalty bidding vs. bonus bidding, etc.). It is often only a matter of opinion as to the "optimal" oil shale leasing method. In a purely economic analysis of potential leasing methods, Jones (1978, p. 23) has determined that the pure bonus bidding system, without royalty or other payments, maximizes economic rents available to the government. Jones believes all other systems are less efficient and reduce the net gains available to society in general. However, other factors (such as social and political considerations) may influence any future leasing decisions.

PERMIT SYSTEMS

Any major oil shale development, like those in coal or other industries, will encounter a maze of regulatory and permit requirements. Rio Blanco Oil Shale Company, the developer of Federal Tract C-a, has identified over fifty permits and approvals required to complete development of a full-scale project. Many of these permits require public notice and hearings which may extend the permit process over many years (Cameron Engineers, 1978a, p. 21).

There is no master plan for meeting regulations or obtaining permits and clearances for oil shale development. Instead, the lessees must determine what regulations affect them and what clearances and permits they must acquire to proceed with development. Often they must even determine what agencies have jurisdiction over their proposed development activities (Novak, 1976, p. 4).

All levels of government (federal, state, county, municipal, and special districts) require permits relating to oil shale development. Davidson (1976, pp. 5-6) reports that the type of data needed by each reviewing agency varies according to the mission of the agency and the purpose of the permit. Often agencies at different levels issue permits for the same element of interest (for example, water quality). In this case, each agency may require almost the same information but in different formats. These situations, of course, compound the paperwork and the time spent by developers in meeting some permit requirements.

Davidson further reports that because it takes years to plan and construct major facilities, companies do not prepare all permit applications to be ready at a single point in time. Generally, permit applications are coordinated with progress in site acquisition and design and engineering. This will occur throughout the five to ten years required for the development of a major project. Of course, each agency receiving an application works within its own time frame for review and approval.

Colorado has published an Environmental Permit Directory (Kinney, 1977) which lists permits required for energy development projects within the State. Table 14.1 lists potential shale regulators at the various government levels, but it must be emphasized that each oil shale site is a specific situation, and one operation may not require the exact set of permits or clearances as another.

It is clear, even from this short review, that permit requirements can be a major constraint to oil shale development. Efforts to simplify and coordinate the process among the various levels of government would greatly assist in the development of oil shale resources.

CAPITAL AND OPERATING COSTS

A "typical" surface oil shale complex consists of four integrated operations: mining and crushing, retorting, waste disposal and shale oil upgrading. Several factors affect the capital investments required to support these operations. These factors include grade of the oil shale and the amount of shale that must be mined and processed to produce a given amount of crude shale oil. Table 14.2 shows the number of tons per day (TPD) of oil shale that must be mined and retorted to produce 104,000 barrels per day (BPD) of crude oil (104,000 BPD of shale oil yields 100,000 BPD of partially-refined syncrude, assuming 96 percent conversion by volume). Generally, both the capital requirements and operating costs decrease significantly as the grade of oil shale increases, so long as crude shale oil production levels remain constant.

In 1974, the Project Independence Task Force developed capital investment and operating cost estimates for six different oil shale complexes. These are displayed in Table 14.3. The estimates show the 50,000 BPD underground and modified <u>in-situ</u> complexes require the smallest total capital investment. However, if the plant size is increased, using the same production technologies, the capital and operating cost per unit of capacity will decrease. In other words, the complex can achieve "economies of scale."

Thus, the smallest capital investment per BPD capacity is the 100,000 BPD for an underground mine with surface retorting. Both the 100,000 BPD underground and the open pit complexes are the least costly, on a unit basis, to operate. The 50,000 BPD underground complex is slightly more expensive, while the <u>in-situ</u> and modified <u>in-situ</u> operations are the least attractive economically. These cost estimates, made in 1974, apparently do not cover the expense for reclamation, vegetation or other environmental protection activities, although they do include estimates for transporting and dumping the spent shale. (See Reclamation and Environmental Costs, Chapter 14.) Of course, because of inflation and cost escalations, the Project

TABLE 14.1

SHALE REGULATORS AND STEPS

P = Permits R = Regulations C = Clearance or Review	Mines & Process	Power Plants	Transmission	Pipelines	Railroads	Roads
FEDERAL GOVERNMENT						
Department of Interior						
Bureau of Indian Affairs	P	P	P	P	P	P
U.S. Geological Survey	P	C	C	C	C	C
Mining Enforcement & Safety	R					
Bureau of Land Management	P	P	P	P	P	P
Bureau of Reclamation	P	P	P	P	P	P
National Park Service	C	C	P	P	P	P
Fish & Wildlife Service	C	C	C	C	C	C
Department of Agriculture						
Forest Service	P		P	P	P	P
Soil Conservation Service	C	C	C	C	C	C
Rural Electric Administration		P	P			
Environmental Protection Agency						
Water Quality	P	P	R	R	R	R
Air Quality	P	P				
Solid Waste Disposal	R	R				
Hazardous Materials	R	R		R		
Department of the Army						
Corps of Engineers	P	P	P	P	P	P
Department of Labor						
Occupational Safety & Health	R	R	R	R	R	R
Department of Transportation						
Federal Aviation Administration	P	P	P			
Federal Highway Administration	C	C	C	C	C	C
Materials Transportation Bureau	R			R	R	
Federal Power Commission		P	P			
Federal Communications Commission	P	P	P	P	P	P
Interstate Commerce Commission				P	P	P
Nuclear Regulatory Commission	P	P				
STATE OF COLORADO						
Department of Health						
Air Pollution Control Div'n.	P	P				
Water Pollution Control Div'n.	P	P				
Engineering & Sanitation Div'n.	P	P				
State Historical Society	C	C	C	C	C	C
Department of Highways						
Division of Highways	P	P	P	P	P	P
Highway Safety Division	R	R	R	R	R	R

TABLE 14.1 (continued)

Department of Labor						
Co. Occupational Safety & Health	R	R	R	R	R	R
Department of Natural Resources						
Div'n of Water Resources (State Eng)	P	P				
Geological Survey	C	C				
Co. Groundwater Commission	P	P				
State Board of Land Commissioners	P	P	P	P	P	P
Division of Mines	R					
Oil & Gas Conservation Comm.	P					
Mined Land Reclamation Section	P					
Co. Soil Conservation Board	C	C	C	C	C	C
Co. Water Conservation Board	C	C	C	C	C	C
Division of Wildlife	C	C	C	C	C	C
Div'n of Parks & Recreation	C	C	C	C	C	C
Public Utilities Commission		P	P	P	P	
Land Use Commission	C	C	C	C	C	C
LOCAL GOVERNMENTS						
County						
Land Use	P	P	P	P	P	P
Air Quality	P	P				
Water Quality	P	P				
Health	R	R				
Fire	R	R	R	R	R	R
Flood	R	R	R	R	R	R
Building Codes	P	P				
Roads	R	R	P	P	P	P
Municipal & Special Districts						
Land Use	P	P	P	P	P	P
Air Quality	P	P				
Water	P	P				
Sanitation	R	R				
Health	R	R				
Fire	R	R	R	R	R	R
Flood	R	R	R	R	R	R
Building Codes	P	P				
Streets	R	R	P	P	P	P

SOURCE: Novak, A., 1976, p. 6.

TABLE 14.2

SHALE TONNAGE REQUIREMENTS FOR SHALE OIL PRODUCTION
(104,000 BPD of Shale Oil)

Grade of Shale (in GPT)	Mine and Retort Requirement* (in tons per day)
25	174,800
30	145,600
35	124,800

SOURCE: National Petroleum Council, 1973, p. 69.

*To produce 104,000 barrels of shale oil per day.

Independence Oil Shale Task Force estimates are not accurate today.

These increased costs can be seen by examining the changing capital cost estimates for the Colony Oil Shale Project. In 1972, the initial capital investment was estimated as $225 million. By March 1974, the estimate had jumped to $425 million, and only five months later reached $653 million (Hale, 1976, p. 50). As of March 1978, the capital costs per daily barrel of shale oil was approximately $20,000, requiring a capital expenditure of one billion dollars for a 50,000 BPD plant. The cost estimates of $20,000 per daily barrel of shale oil compare unfavorably with estimates of $8,000 - $10,000 for North Sea oil and $5,000 - $10,000 for Alaskan oil (Cameron Engineers, 1978a, p. 38).

Table 14.4 shows capital investment and operating cost estimates for the combined in-situ and surface retorting methods. These estimates (Grossman, 1977, pp. 9-15) were made for Tract C-b and did not include lease bonus payments and sunk (see p. 181) costs for environmental background monitoring, engineering studies and preparation of the Detailed Development Plan. This combination of in-situ and surface retorting would yield 54,000 BPD of oil plus propane. Total oil production over a 30-year plant life would be on the order of 500 million barrels. This represents a 55 percent increase in resource recovery over the initial surface retorting only plan (on Tract C-b). This compares well to the estimated 330 million barrels that could be produced solely by room and pillar mining in the Mahogany Zone with surface retorting.

The study also included discounted cash flow calculations at several oil price levels to obtain a curve showing annual rate of return as a function of the value of the shale oil product; all calculations are based on 1976 constant dollars. Shale oil prices, therefore, can be related to the 1976 cost

TABLE 14.3

OIL SHALE COMPLEX CAPITAL AND OPERATING COSTS (1974 Dollars)

MINE TYPE	Undrgnd	Mod IS	Mod IS	In-Situ	Undrgnd	Surface
RETORT TYPE	S	S & U	S & U	U	S	S
CAPACITY (BPD)	50,000	50,000	50,000	50,000	100,000	100,000
SHALE FEED GRADE (GPT)	30	18	25	22	30	30
Capital Investment (millions of dollars)	280	310	280	380	200	520
Operating Cost, excluding depreciation (millions of dollars per year)	45	110	90	140	45	80
Capital Cost per 1000 BPD CAP($)	5600	6200	5600	7600	5200	6000
Operating Cost per 1000 BPD CAP($)	2.46	6.03	4.93	7.67	2.19	2.19

SOURCE: Project Independence Oil Shale Task Force, 1974, p. 65.

Legend: ModIS = modified in-situ
S = surface
U = underground

TABLE 14.4

CAPITAL INVESTMENT AND OPERATING COST FOR COMBINED IN SITU AND SURFACE RETORTING

Capital Investment ($MM, 1976)	
In-situ Retorting:	
Surface plants	8.3
Shafts, tunnels, drifts	56.6
Retort development and operation	58.9
Crushing, ore handling, spent shale disposal	65.4
Surface pyrolysis	141.7
Process facilities	126.6
Utilities and general facilities	120.6
Predevelopment, community assistance and miscellaneous	38.4
	616.5
Deferred mining equipment ($7.7MM/year average)	223.3
	839.8
Operating costs ($MM, 1976)	
Annual cost, at full capacity	69
Initial in-situ demonstration	70

SOURCE: Grossman, 1977, p. 12.

of imported oil and the controlled price of domestic oil. Grossman estimates the necessary value of this low end point shale oil, not hydro-treated, is $12.50 per barrel at the plant site relative to 1976 prices. At this price, the annual rate of return is about 12 percent. By comparison, in-situ retorting alone would yield less than 9 percent return on investment; surface retorting alone, only slightly better than 9 percent return. It appears, from this study, that the combined operation offers an economic advantage. Figure 14.1 shows the results of the study.

However, like other variables in an untested industry, the actual costs of any oil shale facility will be unknown until a plant is actually built and operated. Government encouragement of some type may well be necessary to get the industry off the ground (See Financing and Risk, Chapter 14).

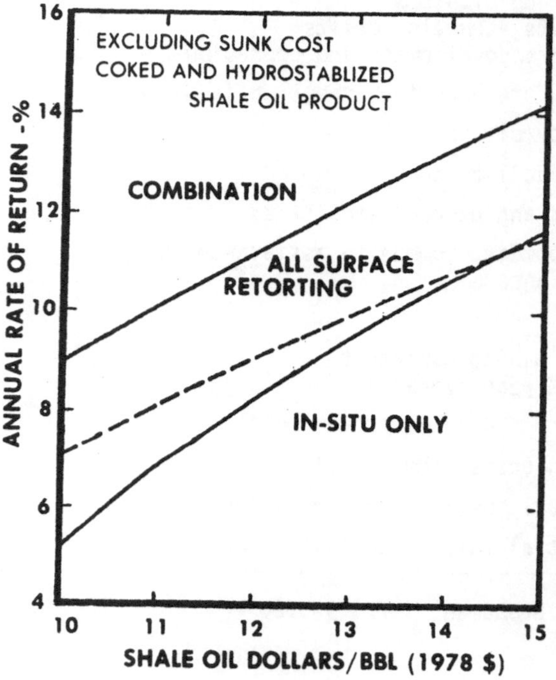

Figure 14.1 Economic analysis by retorting method.
Source: Grossman, 1977, p. 13.

FINANCIAL FACTORS

Economic Uncertainties

Several economic factors presently discourage investment in shale oil facilities. These include:

1. Uncertain shale oil production costs, because of the inflation as well as the difficulty of extrapolating the costs of a commercial plant (more than 50,000 tons per day) from the costs of pilot plant (less than 1,000 TPD) operations.
2. Uncertain environmental protection and reclamation costs.
3. Uncertain governmental levies, such as a "front-end" payment to reduce social impact.
4. Uncertain future quantity and cost of competitive foreign crude oil.
5. Uncertain future quantity and cost of competitive domestic fuels, such as oil, gas, and synfuels derived from materials other than oil shale.

The first three factors listed increase the size of the required capital investments; the last two generate revenue uncertainties. That is, they affect the price the marketplace will bear. Consequently, each factor must be satisfactorily addressed before a sizable commercial shale oil industry will develop.

Taxes

Both potential oil shale development and existing coal mining share similar tax structures (See Coal, Part I, Chapter 5). Major revenues for the State will come in the form of property- income-, sales-, and use-taxes paid by oil shale facilities. Table 14.5 shows projected local and State government revenues from tax sources to be generated over the lifetime of the Tract C-b project. The State severance tax, which took effect January 1, 1978, will generate additional income.

In fact, severance tax legislation will tax oil shale development at a beginning rate of 1 percent of gross proceeds (defined as mine month values) and escalating up to 4 percent of gross proceeds in the fourth and succeeding years of a commercial oil shale activity (50 percent of design capacity). The first 10,000 barrels per day of production will be exempt, with 25 percent credit for the use of <u>in-situ</u> methods (Colorado Mining Association, 1977, p. 184). A portion of this severance tax revenue will be returned to impacted energy development areas through the Local Government Severance Tax Fund (see Socio-economic Impacts chapter).

TABLE 14.5

PROJECTED PROPERTY, INCOME, SALES AND USE TAXES GENERATED BY THE C-b PROJECT BY YEAR FOR A 31 YEAR PROJECT (1975 Dollars)

Phase	Cumulative Years	Rio Blanco Property Tax (on C-b Project) Collected[a]	Total Impacted Counties Property Tax - Individual Collected
Phase I	1	-	-
Mine	2	-	-
Development	3	-	$ 246,924
	4	-	221,948
	5	-	51,522
Phase II	6	-	56,311
Plant	7	$ 228,265	451,971
Construction	8	981,658	1,283,539
	9	2,521,856	2,408,909
Phase III	10	5,133,599	1,948,526
Operation	11	6,568,161	1,299,613
	12	6,568,161	1,405,044
	13	6,568,161	1,382,986
	14	6,568,161	1,382,986
	15	6,568,161	1,382,986
	16	6,568,161	1,420,892
	17	6,568,161	1,448,252
	18	6,568,161	1,459,838
	19	6,568,161	1,459,838
	20	6,568,161	1,459,838
	21-31	6,568,161	1,483,183
Project Life Total		$146,796,759	$37,092,936

<u>SOURCE</u>: Shell, 1976, p. VII-23.

Notes: (a) The difference between the amount assessed and the amount collected reflects a one year lag between property tax assessment and collection. Taxes assessed in one year are collected and are part of the general county revenue the following year.

(b) These calculations are based on a multiplier of 3, i.e. each new "energy dollar" will be turned over three times in the local economy.

TABLE 14.5 (continued)

Phase	Cumulative Years	Income Tax (Wages)	Sales and Use Tax[b] Wages	Sales and Use Tax[b] Non-Wages	Total
Phase I	1	-	-	-	-
Mine	2	$ 602,766	$ 297,837	-	$ 902,603
Development	3	630,696	311,638	-	1,189,258
	4	79,929	39,494	-	341,371
	5	79,929	39,494	-	176,945
Phase II	6	964,689	476,670	$ 60,975	1,558,645
Plant	7	2,751,198	1,359,416	223,575	5,014,425
Construction	8	5,649,765	2,791,648	569,100	11,275,710
	9	4,176,378	2,063,622	1,178,850	12,349,615
Phase III	10	1,522,191	752,142	806,400	10,162,858
Operation	11	1,364,180	674,065	806,400	10,712,419
	12	1,364,180	674,065	806,400	10,817,850
	13	1,364,180	674,065	806,400	10,795,792
	14	1,364,180	674,065	806,400	10,795,792
	15	1,401,909	692,708	806,400	10,852,164
	16	1,429,578	706,380	806,400	10,931,411
	17	1,440,842	711,945	806,400	10,975,600
	18	1,440,842	711,945	806,400	10,987,186
	19	1,440,842	711,945	806,400	10,987,186
	20	1,463,370	723,077	806,400	11,020,846
	21-31	1,463,370	723,077	806,400	11,044,191
Project Life Total		$46,628,714	$23,040,068	$19,773,300	$273,331,777

The fact that the severance tax will not take full effect until the fourth year of activity may be very significant to the development of the industry. Spence (1976, p. 17) reports that oil shale development, an untried industry without proven profitability and facing the risks of any new business, may be discouraged by additional taxation before full-scale production actually begins. The three year grace period for oil shale facilities may allow the industry enough development time before having to pay full severance taxes.

However, Spence also believes that tax schemes which impose taxes on gross proceeds (like the Colorado Severance Tax) rather than on net proceeds ignores the importance of sufficient profitability for the industry. Profit is, of course, the major factor in most business decision-making. Therefore, anticipated taxes before any evaluation of profits may only delay the timing potential of future revenues.

Finally, Spence discusses a fundamental point: tax base is determined by the level of economic activity within the taxing jurisdiction. Business property, private incomes and property values all form the tax base and are created and maintained in part through productive industrial activity. Also, the secondary economic effects resulting from business activity are significant to the tax base. The Colorado Department of Revenue calculates that wages paid to citizens pass through the local economy three times. The mere existence of a business, then, creates and maintains income, property values, and sales taxes through its operations, employees and suppliers.

Whether an oil shale facility will, indeed, pay its "fair share" of taxes to support the impacts on a local economy is an issue to be determined. More evaluation of public costs vs. public benefits of possible oil shale tax structures may be appropriate.

Financing and Risk

Many of the companies which have been most interested in investing in an oil shale industry are the major and independent producers already in the petroleum industry. Since oil companies may have extensive world-wide investment opportunities, oil shale projects must compete for investment funds with domestic and foreign projects, such as conventional on-shore or off-shore oil production. Consequently, the prospects for profitability and economic risk (uncertainty) in oil shale investment must fall within capital investment guidelines used by the oil companies.

A business enterprise may gain capital through internal or external financing. Internal financing is generally secured by retaining earnings, or by selling capital stock or corporate bonds. External financing, in contrast, is usually achieved with a loan from a financial institution; land, facilities, and/or equipment may often be used as collateral. The large financing requirements of a 50,000 BPD oil shale complex--in excess of $1 billion--will no doubt demand significant external financing.

Two factors affecting a lender's willingness to provide capital are the estimated rate of return (ROR) on investments and the "capital exposure" risk. A lender, of course (or investor), wants a "good" return on his investment over a period of time. At the same time, he doesn't want to take an "unreasonable" risk that his funds won't be returned. Rate of return values are quite sensitive to the profit estimates for a project and will affect the amount of capital an investor is willing to provide. Capital exposure, the amount of money a lender stands to lose if the project fails, is sensitive to the level of certainty for payback of the capital invested.

The Synfuels Interagency Task Force (1975, p. IV-E-9) determined that the problems of capital availability and capital exposure are more significant for smaller firms and independent ventures involving several firms. Consequently, the Task Force analyzed several incentives calculated to remove constraints to investment in shale oil ventures.

1. Price guarantees.
2. Tax credits.
3. Loan guarantees.
4. Construction subsidies.
5. Government plant ownership.

Price guarantees help assure profit and thus enhance the availability of capital because of the guaranteed rate of return. However, price guarantees do not reduce capital exposure. In addition, the oil price control enacted by Congress (1975 Energy Policy and Conservation Act) indicates that price guarantees may be politically undesirable, except perhaps in a case of national defense.

Tax credits to encourage investments do not, by themselves, assure profitability. However, tax credits may increase profits and speed the return of capital once an operation reaches a profitable level. As a result, capital exposure is reduced through a more rapid recapture of capital.

A federal loan guarantee program would provide an incentive to invest in a commercial shale oil industry; perhaps the means by which to encourage participation of smaller companies. A federal loan guarantee also signifies the government's commitment to share the risks built into the development of the oil shale industry. The loan guarantee will probably have to be non-recourse; that is, if the participants default on the project they lose their investment capital. However the Federal government is liable for repaying the loan money that private lenders have put into the project.

A non-recourse loan is a very desirable way to finance oil shale development. Because of the technical risks and the large capital requirements, private investors are reluctant to commit funds to the development of prototype shale oil plants. A non-recourse loan, however, not only reduces the capital exposure risk, but may also substantially reduce the break-even selling price of syncrude needed to yield a minimum rate of return. In short, if a company can get a loan to cover some of its oil shale project costs, rather than having to fund it entirely with its own equity (share-holder capital), it can stretch, or leverage, its total funds further (i.e., debt leveraging).

Hoskins and others (1976, p. 7) estimate that a 75 percent loan with a 9 percent interest rate would reduce the syncrude cost by $4.40 per barrel relative to the 100 percent equity case. This is assuming a 10 percent discounted cash flow rate of return (DCFROR). (All capital investments are discounted in relation to the "future value" of the money in order to determine the best future use of company funds.)

In developing the investment prospects for each possible investment, a firm must calculate how many dollars would have to be put in the bank today, at the prevailing interest rate, to yield a given sum at some future date. After knowing what this present-value to future-value discount rate would be, the firm can decide if it would get a better return on its dollars by investing, for example, in an oil shale development, or by simply putting its money in the bank. This and other price-DCFROR combinations are shown in Figure 14.2.

The incentives of construction subsidy and government plant ownership effectively remove the problems of capital availability and capital exposure from the private sector. Unfortunately, these incentives also limit the role of private enterprise in shale oil development, except perhaps as operators.

The Synfuels Task Force (1975, p. I-4) recommends only two of these five incentives: price guarantees and the non-recourse guaranteed loan. The Task Force concluded:

> . . . in the absence of federally provided economic incentives, or other policies creating a stable and favorable investment environment, significant amounts of synthetic fuels are not likely to be produced by 1985.

RECLAMATION AND ENVIRONMENTAL COSTS

If environmental costs such as baseline studies, planning, pollution control engineering, outside consultants, reclamation, etc. are included in capital investment analyses, the impact of oil shale development is significant. For example, in 1975 Hoskins et al., estimated annual reclamation costs at three cents and environmental costs at one cent per ton of oil shale mined. Table 14.6 lists the approximate number of tons of oil shale to be mined annually and the related reclamation and environmental costs for each operation (included in

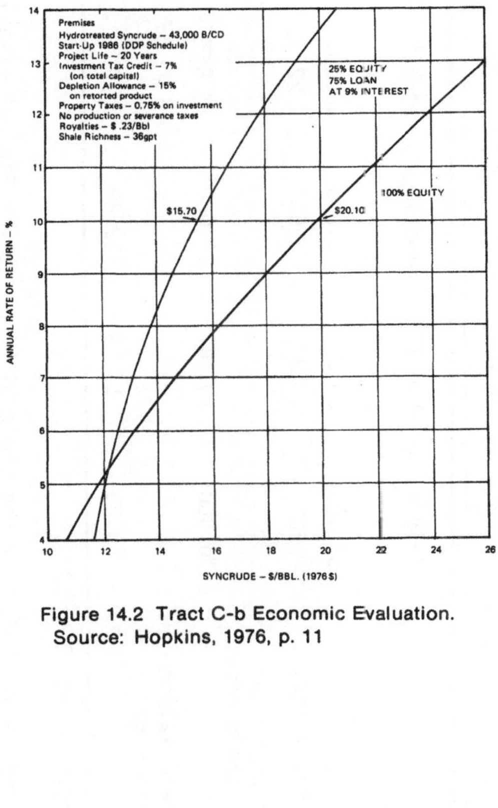

Figure 14.2 Tract C-b Economic Evaluation.
Source: Hopkins, 1976, p. 11

TABLE 14.6

POSSIBLE RECLAMATION AND ENVIRONMENTAL COSTS (1974 Dollars)

MINE TYPE	Undrgnd	Mod IS	Mod IS	In-Situ	Undrgnd	Open Pit
RETORT TYPE	S	S & U	S & U	U	S	S
CAPACITY (BPD)	50,000	50,000	50,000	50,000	100,000	100,000
SHALE FEED GRADE (GPT)	30	18	25	22	30	30
Approximate number of tons of oil shale mined each year*	70,000	29,167†	21,000†	N/A**	140,000	140,000
Estimated reclamation costs (millions of dollars)††	30.66	17.89	12.88	N/A	61.32	61.32

SOURCE: Hoskins, et al., 1975.

*Assuming recovery of 100% of the shale oil content (by Fischer assay) per ton of shale.

**No oil shale is mined; all is retorted in place after drilling and fracturing.

†Only 25% of the oil shale in place is mined; remainder is retorted in-situ.

††Based on a 30-year plant life, with a reclamation cost of $.03 per ton and an environmental cost of $.01 per ton.

Table 14.3). Reclamation and environmental protection costs for in-situ and modified in-situ processes are minimal, since they generate little or no spent shale. However, cost estimates are not available to account for the environmental degradation (such as aquifer contamination or accidental venting of retort products at the surface) which might result from the in-situ retorting process. In addition, in-situ site selection costs may exceed those for more conventional technologies because of the more restrictive geological and hydrological requirements.

Most environmental costs are not included in an economic feasibility analysis because they are considered "sunk costs." A sunk cost, as reported by Stermole (1974, p. 286), occurred sometime in the past and cannot be altered by present or future action. The concept that past costs are not relevant to investment evaluation studies is called the "sunk cost concept." Economic decisions related to future action (such as the building of an oil shale facility) should not be affected adversely by sunk costs.

The most obvious example of an environmental sunk cost is the baseline study consisting of monitoring many environmental variables before development actually beings. (See Environmental Planning, Chapter 15.) Each Federal Oil Shale Tract lessee has spent between five and seven million dollars in the collection of baseline data (Cameron Engineers, 1978a, p. 26). As a result of Colony Development's preparing to build a commercial facility on private land, over 3 million dollars were spent to conduct more than 100 separate environmental studies between 1965 and 1975. The effort included a 20-volume, 6500-page environmental impact analysis describing in detail the existing environment and assessing the possible effects of commercial operation (Cameron Engineers, 1978a, pp. 34-35). While the information generated in this type of study will certainly provide a better understanding of the environmental issues surrounding projected development, the costs of conducting this sort of research represent considerable expense to the companies involved, companies which normally cannot expect a cash return for these environmental study investments.

In summary, there presently exist many obstacles to the commercial development of an oil shale industry. The Federal government has leased only a small fraction of its extensive oil shale land holdings; the amount and type of future leasing is uncertain. To further complicate matters, once an oil shale developer has decided to begin production, he is faced with a complex maze of regulations, permits, and clearances involving considerable delays and duplication of paperwork at different government levels.

Capital and operating costs for oil shale development will be extremely high (one billion dollars or more) for most facilities, and constant cost increases make these cost estimates uncertain. In addition, it appears the cost of producing a barrel of shale oil (including a fair return on investment) cannot yet compete with the free market cost of a barrel of crude oil. Financing for these costly facilities will be hard to obtain because of many economic uncertainties, such as tax levies and their effects on an untested industry, environmental and community impact mitigation costs, and future costs of alternative fuels, to name a few. The commercial development of an oil shale industry will indeed, be a formidable task.

15 Environmental Considerations

One of the major constraints on oil shale development is protection of the natural environment. The environmental impacts associated with the development of a full-scale industry in Colorado could be severe, particularly in the region that will be affected--an area roughly circumscribed by the towns of Rangely, Meeker, Rifle, and Fruita as shown in Figure 15.1--unless steps are taken to mitigate these impacts. At present, the land, air, and water are relatively unspoiled and free from municipal and industrial pollution.

Environmental effects associated with the development of an oil shale complex are characterized as primary (on-site) impacts and secondary (off-site) impacts, although distinction between the two is often artificial or arbitrary (Environmental Protection Agency, 1974, p. 11). Control of the secondary impacts will usually require a regionally-oriented environmental program, since these impacts are widespread, long-term problems affecting the region's ecology, industrial and population growth, recreation resources, and watershed.

This chapter concentrates on the legislative, planning, and disruptive aspects of the primary impacts--that is, site-specific environmental impacts such as air and water quality, surface disturbances, and wildlife habitat disturbances (For a discussion of the socio-economic aspects, see Colorado Oil Shale: Importance and Impact, Chapter 16).

ENVIRONMENTAL LEGISLATION

Federal environmental legislation having the greatest influence on oil shale development includes the National Environmental Policy Act (NEPA) of 1969; the Clean Air Act (CAA) of 1963, amended and expanded in 1967, 1970, and 1977; and the Federal Water Pollution Control Act (FWPCA) of 1948, amended and expanded in 1965 and 1972. A brief discussion of these laws, along with related State laws, is included in Part I - Coal (p. 53). Specific information about these laws and how

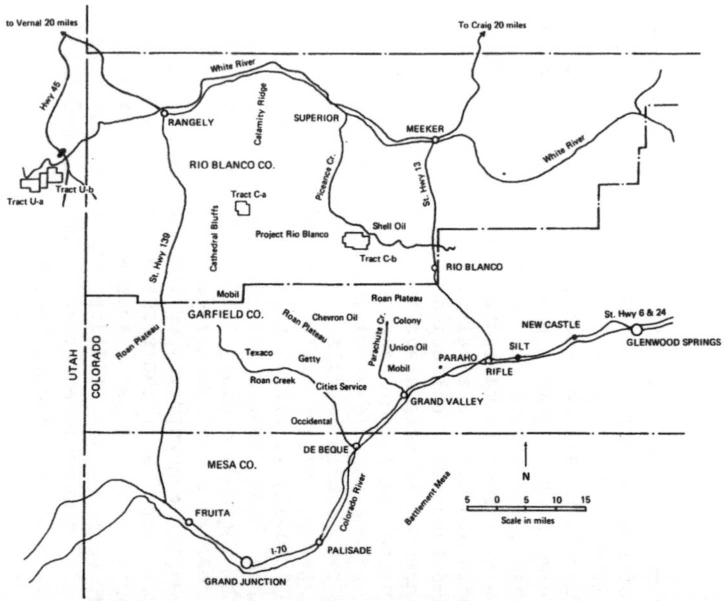

Figure 15.1 Approx. Locations of Oil Shale Environmental Activities
Source: Rocky Mountain Oil and Gas Association, 1975, p. 10.

they may affect oil shale development may be obtained from the Environmental Protection Agency or an appropriate State agency.

ENVIRONMENTAL PLANNING

Potential oil shale developers are engaged in perhaps the most extensive environmental planning program in the history of mineral development. Even before the leasing of federal oil shale lands in 1974, a great many environmental study and planning efforts were active, supported by private oil shale ventures (such as Colony Development Operation) and by government and university researchers (Paxton, 1976, p. 7). Today, each potential operator (on both private or federal land) has gathered an impressive set of environmental baseline data in an effort to develop this resource in an environmentally responsible manner. Cameron Engineers (1978a, pp. 26-28) provides the following brief review of federal lease environment planning.

Under federal lease terms, lessees are required to collect environmental baseline data for a period of at least two consecutive years, one year of which must precede the submitting of the Detailed Development Plan (DDP). These data are used to identify the conditions existing prior to oil shale development on the leased lands in Colorado and Utah. The lessees must conduct monitoring programs before, during, and after development operations and must continue these until the USGS Area Oil Shale Supervisor is satisfied that all federal, state, and local environmental requirements have been met. This environmental assessment program includes evaluation of surface and subsurface hydrology, air quality and meteorology, plant and animal life, soil productivity and archaeology. The environmental baseline data collected, then, serves as the basis for environmental planning.

The procedures and plans for lessening the impact are designed to meet the needs of each development site (as detailed in the DDP). A general outline prepared by the MITRE Corporation (under contract to the Area Oil Shale Office) serves as a model for DDP development by federal lessees. All plans are not expected to conform exactly to this outline.

The basic principles in the outline, in addition to lease requirements, are:

- A development plan which provides an overall picture of the project from beginning operations to final reclamation.
- A plan to have levels of detail which would vary with time--early actions presented in detail, while long range actions would, of necessity, be conceptual.
- A most probable course traced through available alternatives for the life of the lease.

- A discussion of alternatives for major actions where alternatives exist.
- A concise description of environmental impacts and mitigating measures.
- The possibility for revising and modifying approved plans to reflect knowledge gained under the terms of the lease.

Finally, the Cameron Engineers review reports that DDPs have been reviewed and approved for Colorado Tracts C-a and C-b, and are currently undergoing court tests. Approval of a DDP submitted by Utah Tracts U-a and U-b has not been given pending resolution of legal issues concerning the lease.

ENVIRONMENTAL PROBLEMS

Air Quality

The size, location, and number of oil shale processing plants finally built will greatly affect the ability of the industry to meet current or future air quality standards. Air pollution and degradation may come from fugitive dust emissions (particulate matter) as well as from gaseous emissions, including hydrogen sulfide (H_2S), sulfur dioxide (SO_2), nitrogen dioxide (NO_2), carbon monoxide (CO), and various hydrocarbons and particulates, as shown in Table 15.1.

Possible sources of fugitive dust include mining, crushing, retorting, and the disposing of spent shale. The dust produced by these sources may be adequately controlled with water and surface agents; however, the use of water to help control dust emissions places an additional requirement on already scarce water resources in the semi-arid oil shale region. The Office of Technology Assessment (1978, p. 101) estimates that a 50,000 BPD oil shale plant would require from minimal to 4,400 acre-feet of water per day for waste disposal alone, which does not consider other water needs.

The possible gaseous emission sources in oil shale development include retorting, upgrading (partial refining), electric power generating, and operating of gasoline or diesel powered equipment and vehicles. While the oil shale industry will probably meet most State and federal emissions standards, the standards for hydrocarbons and sulfur dioxide pose special problems.

During environmental baseline studies conducted on Tracts C-a and C-b, high background levels of trace hydrocarbon emissions were observed. At times, these background levels even exceeded the National Ambient Air Quality standards established to protect human health. Natural emissions from local flora appear to be the cause of this problem since the land is remote and uninhabited. Oil shale operators on these tracts will have an especially difficult time trying to stay within established environmental protection standards (Paxton, 1976, p. 8).

TABLE 15.1

PREDICTED EMISSION RATES FROM OIL SHALE RETORTING PROCESSES PRODUCING 50,000 bbl/day (tons/year)

Pollutant	In-Situ Retorting	Surface Retorting
Particulates	1,600	300 - 3,250
Sulfur Dioxide	8,500	960 - 5,800
Nitrogen Oxides	2,300	600 - 6,400
Carbon Monoxide	70	300 - 3,600
Hydrocarbons	1,000	1,400 - 4,000

SOURCE: Environmental Protection Agency, 1977, p. 12.

Many steps, however, can be taken by developers to lessen expected air pollution impacts. The Environmental Protection Agency (1977, pp. 11-12) describes typical procedures.

In the crushing and retorting plants, the shale will be carried along in covered conveyors and sprayed with water at points where it is to be transferred from these conveyors to processing units. Baghouse filters, similar to those used in industrial smokestacks, will be used in crusher units to capture the dust from the shale as it is being pulverized. Wet scrubbers will control any particulates generated in retorting operations.

Sulfur emissions from all gaseous fuels produced in the retorting process will first be treated to recover ammonia and will then be desulfurized before being used to heat the shale. Well established techniques will be used to remove sulfur from the retorting and output gases--a chemical recovery system will be used for the low-Btu gases produced in some processes (such as the Paraho process), and high-Btu gases produced in other processes (such as TOSCO process) will be reacted with hydrogen, followed by chemical removal methods, to clean out most of the remaining sulfur.

How well any nitrogen oxides produced in shale oil processing are controlled will depend on how much of the nitrogen in the raw shale can be converted to ammonia before the gases are used to fuel power plants. As an added measure, the burners that use the gas for fuel will be designed to reduce nitrogen oxide emissions. Finally, if liquid fuels are used in shale processing, they will be reacted with hydrogen to transform their nitrogen content to ammonia before burning.

The EPA further states that the amount of carbon monoxide resulting from oil shale operations is expected to be within state and federal standards. If there is any question of the quantity of emissions in the gas streams, the gas will be flared before being released.

Applicable State and federal SO_2 emission standards are shown in Table 15.2. The maximum allowable increases (or increments) above the existing air quality levels limit the amount of air pollutant development in an area to avoid significant deterioration (SD) of air quality. These federal Class II and Colorado Category II incremental limits were designed to allow only a moderate increase in SO_2 levels in certain regions of the State judged to have relatively clean air (U.S. Department of the Interior, 1978, p. 68).

TABLE 15.2

FEDERAL AND COLORADO MAXIMUM ALLOWABLE SO_2 INCREASES (INCREMENTS) FOR THE PREVENTION OF SIGNIFICANT DETERIORATION OF AIR QUALITY (above a specified baseline concentration)

	Averaging Time	Allowed Increase (mg/m^3)
State Standards		
(Category II)	Annual Mean	10
	24 - hour	50
	3 - hour	300
Federal Standards		
(Class II)	Annual Mean	20
	24 - hour	91
	3 - hour	512

SOURCE: U.S. Department of the Interior, 1978, p. 74.

Note: 24- and 3- hour standards cannot be exceeded more than once per year.

According to the Project Independence Oil Shale Task Force (1974, p. 225), four 50,000 barrel per day (BPD) shale oil plants in the Piceance Creek Basin would produce an incremental mean annual SO_2 concentration of 9.9 milligrams per cubic meter based on the removal of 99.5 percent of the sulfur dioxide from an effluent gas. Thus, without changes in the standards or in the efficiency controls, Colorado's mean annual SO_2 standard would limit the Piceance Creek Basin to four 50,000 BPD surface plants.

Likewise, trace elements like arsenic and antimony, are not expected to be emitted from plant gases in large enough quantities to become hazards.

Water Quality and Availability

The possible water pollution sources from oil shale development are: in-situ retorts, retorting plant discharges, drainage from spent shale and flood water. While some of these operations might affect the ground waters underlying the shale region, little data are available about the possible effects. Little is known, for example, about the occurrence and movement of ground water in oil shale beds after in-situ processing. Consequently, there is no information about the amount of ground water, or aquifer, contamination expected from unrecovered hydrocarbons or mineral leaching caused by steam injection. In addition, mine drainage and the use of ground water, with either in-situ or more conventional mines, may lower the pressure of underground water supplies. This could degrade the quality of ground water available in the future.

Current water pollution controls are designed to curb pollutants from entering surface waters. For an oil shale complex, water pollutants could originate from the retorting plant and by leaching from the spent shale. Retorting yields two to five gallons of water per ton of shale--water contaminated with dissolved gases (ammonia and chlorine), dissolved solids (carbonates, sulfates, mercury, selenium and arsenic), and organic compounds (phenols, carboxylic acids) (CSMRI, 1975, p. V. 10). In addition, water used to quench spent shale from the retort, as well as water runoff from a spent-shale disposal area, will pick up silt, salts and organic compounds. To prevent these pollutants from entering natural water courses, oil shale complexes must provide for the retention or impoundment of retort and spent shale water. Upstream diversion dams will also be needed to reduce the probability that these impoundment areas would be breached by flash floods.

There are many existing techniques which may lessen the effects of shale oil development on water quality. The Environmental Protection Agency (1977, pp. 14-15) summarizes these procedures.

Retorting processes being developed incorporate ways of treating wastewater to rid it of contaminants. In a typical plant, wastewater will be cleansed using a series of methods--a foul water stripping system to remove dissolved hydrogen sulfide and ammonia, a holding pond and gravity separation unit to separate solid contaminants, and chemical additives.

Cleaned up wastewater will be recycled through the process stream so that there will be zero discharge of harmful effluents into surface streams and groundwater supplies. Wastewater from mine de-watering, boiler blowdown, and cooling towers will be used directly, without treatment to suppress dust or wet down retorted shale.

All of these measures apply to mining and above-ground retorting. In-situ retorting presents another problem because of the difficulty of controlling the large amounts of wastewater resulting from underground operations. Plans now call for wastewater or process water to be pumped out with oil. A more severe problem may be the soluble salts left in minerals by retorting; these may be picked up by ground water (Hutchins, 1978).

Some disposal methods being considered for in-situ processing include plans to channel the wastewater into the piles of retorted shale left in the ground, to pump it into aquifers that already have high salt concentrations, or to let it collect in large evaporation ponds. However, none of these methods is totally satisfactory, and the problem for in-situ processes remains unsolved.

The EPA further reports that heavy depletion of ground waters increases the salinity of surface streams. The U.S. Department of Interior estimates that a shale oil industry using 120,000 to 190,000 acre-feet of water per year (representing a 1.5 million BPD industry [Hutchins, 1978]) would increase the salinity of the Colorado River downstream at Hoover Dam by 10 - 15 milligrams per liter. The Colorado River is already so heavily salt-laden that desalinization plants are being built on the lower Colorado River. These are necessary to ensure that water obligated to Mexico will be of the quality specified by treaty.

Some groundwater in shale country is as salty as seawater and would be highly corrosive to metal parts if it were used in oil shale operations. Furthermore, saline aquifers disrupted by mining activity could release salt into pure water aquifers adjacent to some of the richest oil shale deposits. As an added problem, saline aquifers punctured by mining activities could flood the mines, thus requiring continuous pumping and disposal of brackish water at the surface in an environmentally acceptable way. At the present time, so little is known about the aquifers and the general groundwater flow patterns that the consequences of oil shale development on ground water cannot be accurately estimated.

Whether water to meet the needs of an oil shale industry is taken from surface or underground sources, there is concern about the possible effects of depletion and the disruption of stream-beds and underground water-bearing strata caused by mining, well-drilling, and spent shale storage. Table 15.3 lists the tentative water availability plans of the major proposed oil shale facilities. Water availability is discussed in some detail in the Nature of the Resource chapter.

Surface Disturbances

The land areas actually affected by oil shale development include those areas required for core drilling, mine development, overburden disposal, storage of low-grade oil shale, construction of surface facilities, and off-site requirements such as access roads, water lines, and gas and oil pipelines. Of course, the amount of surface disturbance will vary, depending upon the type of mining and processing option. In general, in-situ recovery

TABLE 15.3

TENTATIVE WATER SOURCES FOR THE MAJOR POTENTIAL OIL SHALE DEVELOPMENTS

C-b Oil Shale Venture

All process water needed for initial development will be from saline aquifers.

Colony Development Operation

Colony's commercial facilities, including the new community, will consume about 12.5 cfs water (9,000 acre-feet/year) from the upper Colorado River Basin. By comparison, the average flow of the Colorado River past Colony's site is 3.600 cfs. Flows during the 1976-77 drought dropped to 1,300 cfs. Colony's consumption is 0.3% and 1.0% of the average and 1976-77 low flows respectively. This need will be met by direct diversion from the Colorado River at Grand Valley. Colony owns commercial direct diversion water rights on the Colorado River, has purchased option rights on proposed new Colorado River Water Conservation District reservoirs, and is negotiating for supplemental water supplies from existing Bureau of Reclamation Reservoirs. A combination of direct diversion during periods of ample runoff and releases from storage during periods of low river flow will satisfy Colony's water needs without significant impact on agricultural or municipal water consumers in the region.

Rio Blanco Oil Shale Company

(C-a) - All process water from saline aquifer.

Superior Oil Company

All process water from saline aquifers. Potable water for on-site use is discharged.

Union Oil Company

Plans an initial 7,000 barrel daily modular plant with later scale-up approaching 50,000 barrel level. Reserves could support a 150,000 barrel operation. Union has long held rights to Colorado River water purchased from agricultural interests on the Roaring Fork and Parachute Creek drainages. Some of the lands from which Union acquired water rights have been removed from agriculture by urbanization. In approving a change in the diversion point for the water involved, the court decreed that certain irrigated lands affected "shall continue to be assessed by the Garfield County assessor as irrigated lands until such time as said lands may be used for a purpose which produces a higher assessed evaluation."

TOSCO Sand Wash and Paraho

Water supply sources are not yet finalized for either the Paraho or TOSCO Sand Wash project.

SOURCE: Cameron Engineers, 1978a, pp. 12-13.

and retorting requires the least acreage; while open-pit mining with surface retorting requires the most, as shown in Table 15.4.

In comparison to the total land area, surface land disturbance from commercial oil shale development will affect only a small part of land. However, reclamation of these disturbed areas, particularly those used for spent-shale disposal, presents economic and operational problems that have yet to be tested on a commercial scale.

One specific problem relates to the fact that retorted shale becomes greater in volume than unprocessed shale. Depending on the grade of shale being processed, the weight of spent shale is only 80 to 85 percent of the original shale. However, the volume of the spent shale, even after compacting, is at least 12 percent greater than its volume before processing. This happens because, after the mining, crushing and retorting processes, the spent shale includes more empty space.

Under current technology, only 50 to 70 percent of the spent shale produced by an underground mine and surface retort could be returned to the mine (Cook, 1974, p. 30). Consequently, a large part, if not all, of the spent shale will have to be permanently stored on the surface, probably by dumping into canyons or on mesas. Table 15.5 shows the tentative plans for spent shale disposal by the major proposed oil shale developments. Tentative reclamation plans call for the establishment of vegetation directly on the spent shale or on soil placed over the spent shale. This would stabilize the surface of the disposal areas and restore them to some useful purpose. However, the spent shales are too saline for seed germination or plant growth, and are deficient in plant-available nitrogen and phosphorous. Extensive fertilization and irrigation procedures would be needed to support the growth of grazing grasses and shrubs.

While experiments in revegetating shale have been going on for about ten years, the composition of the shale and the lack of plant nutrients in it lead many observers to question the long-term success of revegetation. The uncertainties include concern about plant growth after irrigation and fertilization end; finding the "best" types of plants to revegetate; the possibilities of shale contaminants being carried up from the plant roots and leaves by capillary action bringing salts to the surface and eventually destroying revegetation. Many of these effects will not be understood until more research is completed.

A well-planned and executed revegetation program could result in: (1) increased erosion control; (2) improvement of the natural plant succession toward enduring cover plants; and (3) an increase in the vegetation use as substitute wildlife food and cover (Project Independence, 1974, p. 148). A guide to the ecosystems of the Piceance Creek Basin and to the risks associated with possible disturbance can be found in Tiedeman (1978).

Many reclamation and revegetation experiments related to oil shale development are now in progress. One benefit from the development of revegetation techniques for spent-shale disposal will be that these techniques can be applied to all types of surface disturbances in the semi-arid oil shale region. Colorado State University is one of the leaders in these revegetation experiments. In addition, oil shale developers have spent and are spending considerable time and money studying the problems of revegetation.

Ecological Effects

Ecological systems are involved with oil shale development in two general ways: as a direct recipient of chemical and physical impacts, and as a pathway for dangerous substances to man. Direct impacts on ecosystems are likely to occur, although to what extent or degree is unknown. Disruption of local plant and animal communities may be most severe as a result of waste disposal. Other effects include the disturbances of animals by noise and high levels of human activity and the potential loss in productivity or injury to plants and animals from chronic low-level emissions of such gases as sulfur dioxide and photochemical oxidents (Energy Research and Development Administration, 1977, p. 20).

Surface construction can adversely affect wildlife habitats and migration patterns. A man-made obstacle such as a fence, extended across a mule deer migration path, could prove disastrous for the herds which abound in the oil-shale region. Even if corridors are established to avoid such direct interference, the very presence of man could disrupt or alter the habits and population of many wildlife species in the area (CSMRI, 1975, p. V. 8).

The general pathways by which harmful substances reach man are by inhalation (atmospheric transport), and ingestion (water transport and food chain) (Energy Research and Development Administration, 1977, p. 20). Chappell (1978) is conducting research to determine if oil shale production will lead to emission of significant amounts of toxic trace elements. Five elements are being studied: arsenic, boron, fluorine, molybdenum and selenium. All of these, except arsenic, are essential to the life of plants, animals, or both; however, at sufficiently high levels, all can be toxic.

Additionally, concerns have been expressed about the potential exposure to cancer-causing elements in oil shale rock and rock from which shale oil had been extracted. Coomes (1978) reported recently on tests performed by TOSCO Corporation on mice and hamsters exposed to those materials by skin, ingestion, or lung exposure. The experimental data showed toxicities equal to or less than those observed through similar contact with petroleum refinery products.

TABLE 15.4

LAND REQUIREMENTS FOR OIL SHALE PROCESSING

Source of Requirement	Land Required (Acres)
Surface Mine[1,2] (100,000 bbl/day):	
Mine development	30 to 85 per year
Permanent disposal; overburden	1,000 (total)
Temporary storage; low-grade shale	100 to 200 (total)
Permanent disposal; processed shale	140 to 150 per year
Surface facilities[3]	200 (total)
Off-site requirements[4]	180 to 600 (total)
Underground Mine[2] (50,000 bbl/day):	
Mine development (surface facilities)	10 (total)
Permanent disposal:	
All processed shale on surface	70 to 75 per year
60-percent return of processed shale underground	28 to 30 per year
Surface facilities[3]	140 (total)
Off-site requirements	180 to 225 (total)
In situ processing (50,000 bbl/day):	
Surface facilities[3]	50 (total)
Active well area and restoration area	110 to 900
Off-site requirements	180 to 600 (total)

SOURCE: Project Independence Oil Shale Task Force, 1974, p. 139.

[1]Area required is dependent upon the thicknesses of the overburden and oil shale at the site. Acres shown are for a Piceance Creek Basin site, with 550 ft of overburden and 450 ft of 30 gallon/ton shale (approximately 900,000 bbl/acre).

[2]Assumes 30 gallons per ton oil shale and a disposal height of 250 ft.

[3]Facilities include shale crushing, storage and retorting (excluded for in situ processing), oil upgrading and storage, and related parking, office, and shop facilities.

[4]Includes access roads, power and transmission facilities, water lines, natural gas and oil pipelines; actual requirements depend on site location. A 60-ft right-of-way for roads requires a surface area of about 8 acres per mile. Utility and pipeline corridors 20 ft in width require 2.4 acres per mile.

TABLE 15.5

TENTATIVE SPENT SHALE DISPOSAL PLANS FOR THE MAJOR POTENTIAL OIL SHALE DEVELOPMENTS

C-b Oil Shale Venture
Use of modified in situ technology will minimize shale disposal. Mined shale resulting from retort development will initially be disposed of on the surface. The disposal pile will be contoured and compacted, followed by covering with a layer of top soil and revegetation. Consideration is being given to returning mined shale to abandoned portions of mine at a future date.
Colony Development Operation
Shale for this operation would be mined from adits in Parachute Creek Canyon and transported to the plateau above, where the plant site would be located. Spent shale disposal would be in the valley atop the plateau with compaction and topsoil overlay prior to revegetation.
Rio Blanco Oil Shale Company (C-a)
Both modified in situ and surface retorts will be utilized. Spent shale from surface retorts will be disposed of on the surface by compaction, contouring, covering with soil, and vegetation. Studies are underway to evaluate back filling of burned out retorts with a spent shale slurry.
Superior Oil Company
All spent shale from this deep mining operation will be returned underground for additional roof-and-pillar support. Because of the multiple minerals in place and recovery operations planned, the volume of the spent shale will be less than that of the raw shale coming from the mine.
Union Oil Company
Shale will be mined from adits in the canyon wall with processing in plants built on a canyon shelf. Spent shale disposal will be in the East Fork Canyon with direct revegetation of the spent shale.
White River Shale Project (U-a and U-b)
Processed shale from surface retorts will be disposed of in a canyon on the tract. The spent shale will be contoured and revegated after the disposal area has reached its final elevation.
TOSCO Sand Wash Paraho
Detailed plans have not yet been developed for spent shale disposal on either the TOSCO Sand Wash or Paraho projects, but it will either be deposited in canyons on the surface or returned underground to mined-out areas.

SOURCE: Cameron Engineers, 1978a, pp. 10-11.

Summary

An oil shale complex will involve retorting, refining, and, except for "pure" in-situ recovery operations, mining, crushing and waste disposal; each of these operations presents possible environmental hazards. Table 15.6 shows a physical environmental impact summary of oil shale development. The hazards of development could be significant; their management will require diligent efforts by both developers and government administrators. The pressure for decisions will grow as oil shale development approaches commercial feasibility; however, it is clear we do not fully understand the long-range environmental effects of a large-scale oil shale industry.

TABLE 15.6

PHYSICAL ENVIRONMENTAL IMPACT (Summary)

1 Barrel of Oil Shale:	
Will use:	2 - 6 barrels of water
	1 - 2 tons of oil shale
	100 square feet of land
Will produce:	1/3 lbs of airborne dust
	2-1/2 lbs of polluting gases
	2 - 5 gallons of contaminated water
	1 - 1-1/2 tons of spent shale

SOURCE: Environmental Protection Agency, 1977, p. 19.

16 Colorado Oil Shale: Importance and Impact

EMPLOYMENT, MANPOWER AND TRAINING

Manpower requirements for an oil shale industry vary according to the phase of development, type of mine operation and retort process, and the design production capacity of the oil shale complex. In general, an oil shale development complex will involve three phases: pre-production activities, production operations, and terminal or shut-down actions. Pre-production activities include mine development, site preparation, and construction and preparation of required facilities and utilities. Production operations include recovery and processing of oil shale, waste disposal, and upgrading of shale oil. Terminal actions include site restoration and salvage operations.

The pre-production activities, which normally span a period of four to eight years, generate manpower requirement "peaks" which exceed the manpower requirements for any subsequent phases. Table 16.1 shows the manpower requirements for the construction of two different oil shale plants, as estimated by the Synfuels Task Force (1975, p. III-C-5 and 6). The peak manpower requirement occurs in the third year of construction, when construction activity is highest.

Once an oil shale complex is producing at design capacity, manpower requirements stabilize at about one-third the level of the peak requirements. Table 16.2 lists the annual manpower requirements for five different oil shale complexes operating at design capacity (Project Independence, 1974, Appendix H). The pure in-situ operation calls for the smallest operating force, in terms of total requirements. The modified in-situ operation requires the next-smallest force.

Economies of scale can reduce the net labor requirement of an oil shale complex for each increment of 1,000 BPD of capacity. For example, the manpower requirement for a 100,000 BPD underground operation is 17 percent less than that required for two 50,000 BPD operations. In fact, the net manpower requirements, in man-years per 1,000 BPD capacity, for the 100,000 BPD open pit operation is less than that for the 50,000 BPD modified in-situ plant (18.0 versus 23.1) (Data were not available on the manpower requirements estimated for a 100,000 BPD modified in-situ operation).

Manpower data on the establishment of a 50,000 BPD underground mine and surface retort complex were recently made available in Shell and Ashland Oil planning documents (Shell, 1976, p. II-19) as shown in Table 16.3 and Figure 16.1. Note that the manpower requirements during the peak construction period are considerably higher than similar estimates made in 1974 (Cf. Table 16.2), but operating manpower levels are lower.

Manpower requirement estimates by phase of development for the Rio Blanco Shale Project on Tract C-a (Rio Blanco Oil Shale Project, 1976) are similar to these estimates made by Shell for Tract C-b, and are shown in Table 16.4. Overall employment for this planned development called for employment to peak at 2,700 during the period when both initial processing to obtain transportable shale oil (9,000 barrels per day) and construction of the full scale facility (which would produce 55,800 barrels per day of upgraded shale oil) were underway.

It is unlikely that all seven operations planned for the Piceance Creek Basin would begin commercial development at the same time let alone begin plant construction simultaneously. However, it is apparent that the manpower requirements to support construction and operation of even a 50,000 BPD plant could severely strain the manpower resources in the oil shale region. Data from other studies indicate that oil shale developers such as Rio Blanco Oil Shale Project, plan to work with federal, State, and local governments to develop recruitment and training programs with emphasis on recruiting and training local residents first. There are also plans for scheduling and maintaining manpower requirements at consistent levels to minimize population expansion and contraction.

SOCIO-ECONOMIC IMPACTS

Colorado's oil shale region is located in a very sparsely settled region on the western slope of the Rocky Mountains. The shale deposits themselves are concentrated in the Parachute Creek area bounded by the small towns of Rangely, Meeker, Rifle, and Grand Valley (See Fig. 15.1). The larger, resort community of Glenwood Springs is approximately 75 miles east of the Parachute Creek area; Grand Junction, the region's major trade and services center, is approximately 110 miles from the C-b lease site.

As of 1977, approximately 90,600 persons lived in the oil shale country of Colorado (Environmental Protection Agency, 1977, p. 22). Leading industries in the region are agriculture, ranching, and tourism, with substantial contributions from both energy production (coal, oil, and gas) and manufacturing. Rapid growth (mostly from energy development, tourism, and recreation)

TABLE 16.1

OIL SHALE PLANT MANPOWER REQUIREMENTS (Man Years)

	50,000 BPD Modified In-Situ Operation				
Year of Construction	1	2	3	4	Total
CONSTRUCTION CRAFTSMEN	0.0	277.0	1284.1	1827.0	3388.1
ENGINEERS AND OTHER PROFESSIONALS	75.7	18.5	49.0	49.9	193.1
TECHNICIANS	92.9	49.6	49.1	34.2	225.8
ADMINISTRATIVE WORKERS	5.3	5.6	7.6	3.5	22.0
TOTAL MAN YEARS (2080 hrs/man-yr)	173.9	350.7	1389.8	1914.6	3829.0
	50,000 BPD Underground Mine and Surface Retort				
Year of Construction	1	2	3	4	Total
CONSTRUCTION CRAFTSMEN	0.0	273.0	1220.0	652.0	2145.0
ENGINEERS AND OTHER PROFESSIONALS	70.1	4.9	16.4	4.4	95.8
TECHNICIANS	80.7	46.5	37.7	7.9	172.8
ADMINISTRATIVE WORKERS	5.1	4.2	8.3	1.8	19.4
TOTAL MAN YEARS (2080 hrs/man-yr)	155.9	328.6	1282.4	666.1	2433.0

SOURCE: Synfuels Task Force, 1975, pp. III-C-14 and 15.

TABLE 16.2

ANNUAL OPERATING MANPOWER REQUIREMENTS[1] (Man-Years per year of operation[2])

Production Rate (BPD)	50,000	50,000	50,000	100,000	100,000
Type of Mine/Retort	In-Situ	Modified In-Situ	UGM SR	UGM SR	OPM SR
PERSONNEL					
Engineers	11	12	13	17	17
Other Technical Personnel	68	64	76	131	121
Managers and Administrators	11	18	25	51	36
Sales Workers	4	4	3	6	6
Clerical Workers	17	21	15	26	22
Construction Craftsmen	148	196	228	442	374
Foremen	36	67	81	141	113
Metal Working Craftsmen	1	1	8	17	17
Mechanics	17	125	218	334	171
Other Craftsmen	0	82	0	0	8
Operators (excluding transport)	244	403	511	810	629
Transport Equipment Operators	0	113	164	227	130
Service and Protective Workers	47	34	80	143	142
Laborers	17	17	8	17	17
TOTAL	621	1157	1430	2362	1803
AVERAGE REQUIREMENT (man-years per 1000 BPD capacity)	12.4	23.1	28.6	23.6	18.0

SOURCE: Project Independence, 1974, Appendix H.

Legend: UGM = Underground Mine
SR = Surface Retort
OPM = Open Pit Mine

[1] Includes manpower requirement for operation and maintenance.

[2] One man-year - 2080 hours.

TABLE 16.3

PROJECTED EMPLOYMENT FOR THE TRACT C-b SHALE OIL FACILITY

Development Phase	Project Year	Total Construction Force	Total Operations Personnel	Total Basic C-b Employment
	1	--	--	--
Mine	2	370	55	425
Development	3	328	62	390
(5 years)	4	--	59	59
	5	--	59	59
Plant	6	672	103	775
Construction	7	1,717	233	1,950
(4 years)	8	3,015	310	3,325
	9	2,670	370	3,040
	10	451	799	1,250
	11-14	0	1,093	1,093
Production	15	0	1,123	1,123
Operations	16	0	1,145	1,145
(22 years)	17-19	0	1,154	1,154
	20-31	0	1,172	1,172

<u>SOURCE</u>: Shell Oil, 1976, p. II-19.

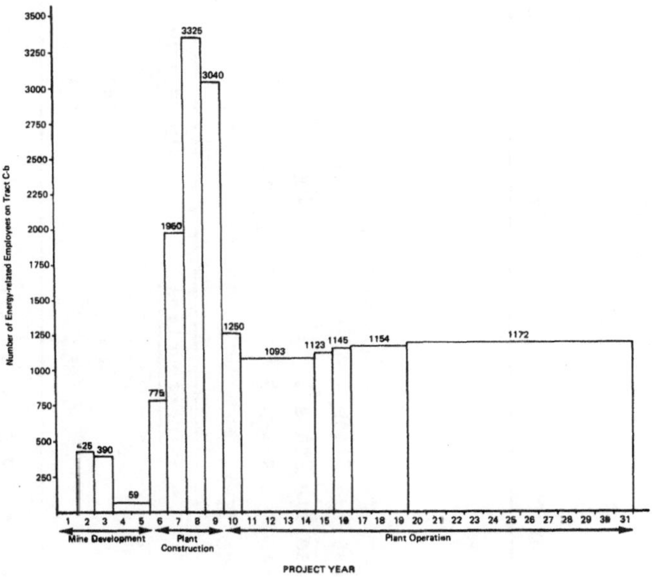

Figure 16.1 Projected On-site Employment for Tract C-b
Source: Shell Oil, 1976, p. II-19.

TABLE 16.4

RBOSP CONSTRUCTION, OPERATION AND EMPLOYMENT SCHEDULE

Activity		Peak Employment		
Construction	Operation	Construction	Operation	Total
Access Road & Power Line	-	300	-	300
Phase I - Stage 1 Product Pipeline	-	700	-	700
-	Phase I - Stage 1	-	300	300
Phase I - Stage 2	Phase I - Stage 1	400	300	700
Phase II	Phase I - Stage 2	2,200	500	2,700
-	Phase II	-	1,100	1,100

Employee Categories	Construction	Operations
Administrative, Management & Technical	15%	24%
Hourly	85%	76%

SOURCE: Rio Blanco Oil Shale Project, 1976, p. 2-1-11.

Note: These estimates do not include manpower requirements for site preparation or mine development; similar construction manpower data were not available for in-situ or open-pit projects.

is already taking place. Population has increased 158 percent from 1970 to 1977 in DeBeque, 96 percent in Silt, and 85 percent in Grand Valley (Federal Energy Administration, 1977, p. 36).

Comparing this type of growth with optimal growth rates puts it in perspective with energy relate growth.

- Some sociologists believe an annual population growth rate of 5 percent is the maximum rate tolerable for the timely provision of housing, schools, health facilities and other necessary community services.
- Growth rates as high as 7 and 10 percent cause boomtowns, placing stress on both services and people.
- A growth rate beyond 10 percent may cause local government structures to falter and to fail to accommodate the new populations, often resulting in chaotic conditions (Environmental Protection Agency, 1977, p. 21).

When forecasting the socio-economic impacts (both positive and negative) of energy development, two issues to consider are the timing (phase-in) and the magnitude of these industrial developments. At this time, there is simply not enough information to accurately predict how much oil shale development will take place in Colorado, or over what time period. This chapter must be limited to an outline of the major socio-economic impacts expected to accompany this development. However, detailed data provided on the three-county area of Mesa, Garfield, and Rio Blanco counties is used where it is appropriate.

The major socio-economic impacts associated with commercial development of an oil shale industry will be: (1) the population growth resulting from the (new) work force required for construction and operation of an oil shale complex; (2) the secondary (induced) population growth associated with providing goods and services to the energy-related workers and their families; (3) the demand for housing and municipal services (education, police, fire, recreation, medical, etc.) generated by this growth; (4) the impacts on the local and regional economies, including new public resources and expenditures, greater selection of available goods, new employment opportunities; and (5) increases in social problems such as alcoholism, divorce, crime, hyperurbanization (increased traffic, congestion, etc.), and more.

A detailed description of the employment and population build-up associated with the development of a 50,000 BPD plant on Tract C-b has been projected by Shell (1976, p. II-19), and is shown in Table 16.5. The proposed operation is approximately forty to fifty miles from the towns of Meeker and Rifle. Related to this development are the population data displayed in Figure 16.2. These figures compare the population projections for the town of Rifle in 1980 with the population anticipated with Tract C-b development during (1) the peak year of

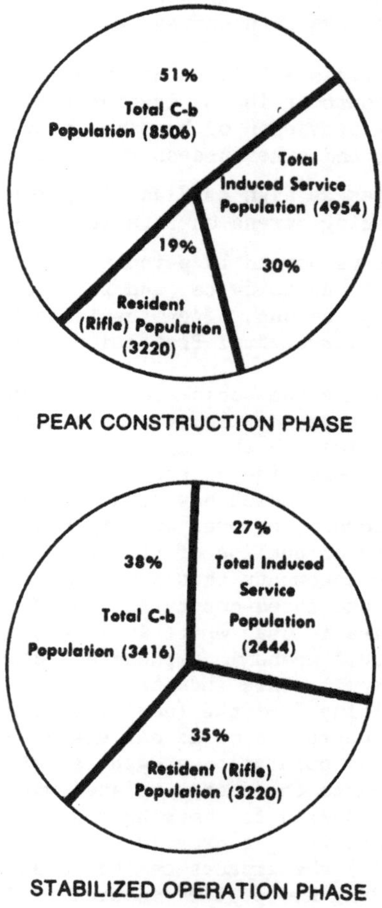

Figure 16.2 Peak Construction and Stabilized Operation Phases
Source: Text.

TABLE 16.5

PROJECTED EMPLOYMENT AND POPULATION BUILD-UP FOR THE SHELL/ASHLAND TRACT C-b SHALE OIL FACILITY

Development Phase	Project Year(s)	Total C-b Employment	Total C-b Population	Service Employment Multiplier	Total Induced Service Employment	Total Induced Service Population	Total Employ-ment	Total Popu-lation
	1	---	---	---	---	---	---	---
Phase I	2	425	1,097	.5	212	316	637	1,413
Mine	3	390	1,014	.5	195	291	585	1,304
Development	4	59	184	.5	30	45	89	229
(5 years)	5	59	184	.5	30	45	89	229
Phase II	6	775	2,002	.5	389	577	1,163	2,579
Plant	7	1,950	5,021	.75	1,473	2,179	3,413	7,200
Construction	8	3,325	8,506	1.0	3.325	4,954	6,650	13,461
(4 years)	9	3,000	7,831	1.0	3,040	4,530	6,080	12,361
	10	1,250	3,624	1.25	1,563	2,328	2,813	5,953
Phase III	11-14	1,093	3,416	1.50	1,640	2,444	2,733	5,860
Production	15	1,123	3,507	1.50	1,684	2,511	2,807	6,018
Operations	16	1,145	3,578	1.50	1,718	2,559	2,863	6,137
(22 years)	17-19	1,154	3,606	1.50	1,731	2,579	2,885	6,185
	20-31	1,172	3,662	1.50	1,758	2,618	2,930	6,280

SOURCE: Shell, 1976, p. II-19.

construction and (2) a representative year of operation. It is interesting to note that in the peak year of construction, the town's residents are outnumbered more than four-to-one by the total shale-related population. Once production operations begin and the total C-b employment stabilizes there is only a two-to-one increase in population. However, this example does not necessarily represent potential population impacts since it is highly unlikely that the town of Rifle would bear the full brunt of growth related to C-b development.

A rough picture of possible oil shale-related population growth can be drawn by applying the same factors as used by Shell/Ashland for Tract C-b as shown in Table 16.1. For example, 4,615 energy employees are predicted for the region in 1985 if six oil shale plants are developed, excluding the development of Colony's private plant. This increase would result in an energy-related population of 14,399 (based on a total of 3.12 persons for each person's family) and an induced service population of 10,308 (using a basic employment to service employment multiplier of 1.5 and a factor of 1.489 for conversion of service employment to service population). Consequently, the total regional population increase in 1985 would be 24,707 without the Colony development, and 30,274 if Colony's development is included. The basic resident population for the three-county area in 1985, without oil shale development, is estimated at 133,900 persons (See Shell, 1976, p. II-21 for details).

As the oil shale-related population grows, the region's need for housing, schools, roads, and all other government services will also increase. The rate of expanding service needs can generally be divided into two phases: Phase I covers the years in which a construction work force dominates the employment scene, while Phase II covers the years in which a more stable operating force is employed. The construction work force is generally considered to be highly mobile, and is made up of a higher proportion of single men. Those married workers who do bring their families tend to live in mobile homes, and do not place demands on the community for permanent housing and permanent school facilities. Nonetheless, construction workers' mobile homes are very likely to create sprawling mobile home communities with possible health hazards and very little opportunity for any community integration.

Perhaps the most serious impact of rapid growth is the greater demand for municipal and social services--police and fire protection, medical services, educational services, public transportation, and others. The key to the problem, of course, is financing: how can the local communities provide necessary services when they are needed (not lagging far behind population growth)?

For many smaller communities in the oil shale region, annual revenues for municipal operations barely cover annual operating expenses. Consequently, most capital improvement expenditures must be financed by municipal bond issues, which, in most cases, are tied to statutory bonding limits based on assessed property values.

Thus, even if growth and the increased demand for services is seen as permanent, and not simply as a transient surge, securing the funds to build additional public facilities will be extremely difficult. Besides the difficulty securing capital financing, knowing how much money is needed is also a problem. Per capita one-time, front-end capital costs will vary according to the development patterns of incoming population. For example, in the Tract C-b area, the most likely settlement pattern alternatives for the shale-related population appear to be: (1) expansion of the existing towns of Meeker and Rifle; (2) low density sprawl along the road between Meeker and Rifle; or (3) the development of a new town at a location near the Rio Blanco Store or some other convenient "midpoint." Land use, zoning, personal taste, economics, environmental, and social factors will combine to determine the final settlement pattern.

The estimated per capita costs associated with each development pattern are shown in Table 16.6. Managed growth--expansion of existing towns--is the least costly; development of a new town is 15 percent more costly than managed growth; and low density sprawl is the most costly--more than 30 percent more than the managed growth alternative. Thus effective planning can reduce the capital cost of rapid, large growth. Planning costs, however, are not included in cost projections.

Though new growth increases public service and facility requirements, it carries the benefits of increased revenues for both the local and state government. The revenues for local governments (municipality, school district, and county) come from _ad valorem_ and sales taxes (See Taxes, Chapter 14), fees, and intergovernmental transfers. Sales taxes and fees (for example, sewage hookup and building permit fees), collected promptly, are revenues responsive to the increased costs of services. In contrast, the collection of _ad valorem_ taxes, in particular property taxes, may lag from one to two years behind the demand for services. Intergovernmental transfers (including federal revenue sharing, state aid to school districts, state highway users' tax revenues, special motor vehicle registration fees, welfare grants, etc.) are more responsive to the immediate needs for community services.

Major State revenue sources linked with oil shale development include income and sales taxes; revenues from the Federal Mined Land Leasing Act of 1920; the Oil Shale Trust Fund; interest earned on the Oil Shale Trust Fund; and the State Severance Tax. The largest source of funds available to the state for resource development impacts is the Oil Shale Trust Fund, which is formed from oil shale lease payments. To date, the state has received three of the five bonus lease payments on oil shale Tracts C-a and C-b, amounting to approximately $74 million. The remaining two lease payments are expected to be credited to on-site development, and the Oil Shale Trust Fund will probably not grow larger until further leasing activity begins (Colorado Socio-Economic Impact Office, 1978, pp. 14-16).

TABLE 16.6

TOTAL PER CAPITA ESTIMATED COSTS ASSOCIATED WITH THREE DEVELOPMENT PATTERNS

Type of Development	Managed Growth (Expansion of Existing Towns)	New Town	Low Density Sprawl (no planning)
Municipal Utilities	1,830	4,480	2,590
Municipal streets & roads, water, sewer, solid waste, cables, storm drainage			
Public Schools	1,574	1,360	1,574
Vocational-Technical	542	542	542
College	199	199	199
Nursery	111	111	111
Other Human Services	572	860	572
Elderly Services	708	708	708
Growth Planning & Consultants	215	140	---
Sub-total	5,751	8,400	6,296
Residential Housing Costs	9,497	9,497	14,246
Sub-total	15,248	17,897	20,542
State and County Highways	2,246	2,246	2,246
Grand Total	17,494	20,143	22,788

SOURCE: Shell, 1976, p. I-17

The Department of Local Affairs, the Governor's Socio-Economic Impact Office, and the Colorado Energy Impact Assistance Plan have additional information on State revenue sources.

To aid in financing the needs of impacted communities, Congress is considering a Federal Socio-Economic Impact Aid Program. Financial aid in this form would help to ease the front-end financing problems facing most energy communities through low-interest loans and loan guarantees as development begins.

In summary, by 1985 even moderate oil shale development could generate serious socio-economic impacts in the Colorado oil shale region of Garfield, Mesa, and Rio Blanco counties. The commercial operation of a single 50,000 BPD oil shale plant would increase the local population by approximately 6,000 persons (3,500 plant employees and dependents; 2,500 induced service workers and dependents). Since the majority of existing communities have populations under 4,000--with several having fewer than 500--the growth rate associated with oil shale development should substantially exceed the historic rate of 3 - 5 percent.

Extremely rapid growth may degrade the existing quality of life by changing settlement patterns, raising the cost of living, and creating dependence upon a specialized industry. Rapid, uncontrolled growth also challenges the agility of local governments to provide municipal and human services at levels required for a healthy social environment. Since the local government revenues needed to provide services lag one to two years behind the demand for increased services, financial assistance from the State or Federal government will continue to be needed in many cases where growth exceeds the optimal growth rate of 5 percent.

SAFETY AND HEALTH

There is the risk of physical injury to workers involved in oil shale mining and processing, just as there is risk in any other major mining or refinery operation. The risk, however, can be reduced considerably through proper procedures and strict industry compliance with health and safety regulations governing these operations (Environmental Protection Agency, 1977, p. 20).

Safety techniques in both underground and surface oil shale facilities are similar to those used in coal mines. However, most of the shoring techniques required in underground coal mining will probably not be required for oil shale mining because the shale's greater strength and stability will provide a more stable mine roof. Nonetheless, both expanding-head and epoxy roof bolts provide an additional safety margin. Since tunnels and rooms can be much larger in underground oil shale mines than in underground coal mines, ventilation should not be a major problem and should reduce, if not eliminate, the danger of toxic gases being trapped in areas of mining (University of Oklahoma, 1975, p. 2-13).

One of the major health concerns for the work force in an oil shale processing facility is exposure to toxic materials. Trace quantities of such materials are either present in oil shale or are produced chemically during retorting and upgrading. Toxic materials may be present in fugitive emissions, liquid and gas streams, waste waters, or processed shale. The health hazards associated with physical contact with these toxic materials need to be evaluated scientifically in relation to realistic doses, or exposures (Energy Research and Development Administration, 1977, pp. 19-20). For an example of the type of study being done to evaluate the health impacts of oil shale workers, see Costello (1978, pp. 1-10) and Coomes (1978, pp. 26-1-26-29).

Besides the health hazards to oil shale workers, effects on nearby populations must also be considered. Any of the various pollutants from oil shale facilities (see Environmental chapter) are either known or suspected to be toxic to man. Many of the biological assays, dose-response data, etc., used to quantify possible risks to the occupational work force will be used in evaluating risks to the general population (Energy Research and Development Administration, 1977, p. 20).

Careful environmental planning can reduce the quantity of these pollutants, but will not eliminate them altogether. Little is known about the ways in which pollutants are carried through the environment to man, or about how the human body resists or adapts to different concentrations of pollutants, and no realistic estimate can be made of how these will ultimately affect the ecology of the surrounding environment and the health of its populations (Environmental Protection Agency, 1977, p. 20).

17
Shale Oil Production Constraints

Many factors affect both the time period and the scale of an oil shale industry. The principal impediments to developing 300,000 barrels per day oil shale industry by the mid-1980s are economic, technological, and environmental factors and their related socio-economic impacts.

ECONOMIC FACTORS

The price together with capital availability will remain uncertain. This is particularly so since large investments are required for such high-risk ventures.

TECHNOLOGICAL FACTORS

While modified <u>in-situ</u> technology is close to supporting commercialization, the technology is still in the experimental stage. Additionally, while current open-pit or underground mining technology can be transferred to the oil shale industry, such massive quantities of material must be removed and processed that slight miscalculations in design or operation could result in significant unnecessary costs.

ENVIRONMENTAL FACTORS

Some of the environmental problems facing commercial oil shale development include air and water pollution, surface and ecological disturbance, disposal of spent shale, reclamation and water availability. Careful planning, based on extensive information gained from baseline studies of the area, may lessen these impacts. However, it is uncertain whether all current federal and State rules, regulations and standards can be met if a commercial oil shale industry is to be developed.

Perhaps the most serious obstacle to development will be the air pollution regulations. Both high natural background emission levels in oil shale areas and uncertain control technology severely limit new development. In fact, it is entirely possible that these environmental constraints may limit development to only a few full scale facilities in the Piceance Creek Basin.

SOCIO-ECONOMIC FACTORS

Socio-economic impacts include population settlement problems ("urban sprawl"), quality of life degradation, and problems associated with providing municipal and county services. While no simple solution is available to eliminate such impacts, the most serious "growth pains" can be reduced by preplanning at all levels of government and by responsible financial assistance by the Federal and State governments.

18 The Future: Observations and Concerns

The first demonstrated recovery of oil from shale in the United States came in 1855 when a retort was constructed and operated near Juab, Utah. Since that time oil shale development in this country has been a history of false starts. Several times, shale has been pronounced "on the verge" of commercial development, but somehow commercialization has never been achieved. Today, the story is not much different: once again this resource has come to the "verge" of commercial development, and once again commercialization may be put off for the present.

SHALE OIL COMMERCIALIZATION

Many obstacles exist today in developing a commercial industry. The technology for underground and surface mining, with surface processing, and in-situ technologies, although growing, remains limited and no commercial processing has been performed thus far. At this time there are few economic incentives to develop and operate commercial-size equipment. Thus, forecasting the time-table for oil shale development is tenuous at best.

There are indications of commercial plants operating in the mid-1980s. However, the current pilot-stage technologies, other than those of Paraho and TOSCO, will probably need more refinement before being scaled-up for a commercial-size operation (Hutchins, 1978). And it is seriously questioned if enough experience has been gained to make design and operational improvements within this relatively short time span.

Research and Development Efforts

The pace of oil shale commercialization will certainly be affected by the extent of government-supported research and development of surface and/or in-situ retorting technologies. The research on technologies must be accompanied by research on related environmental factors, including possible contamination of ground water, subsurface movement of by-products and air pollution resulting from processing. Intensive research of both technological and environmental concerns can certainly speed the commercialization of shale oil recovery.

Commercial development of oil shale will also depend on economic factors. Federal loan guarantees for commercialization of synthetic fuels would speed the development of oil shale. Other economic alternatives may include: accelerated cost depreciation; investment tax credits for the plants; enhancement of capital formation (S-419, F. Haskell, D. Colo.); or some combination of these incentives.

When commercial oil shale development is finally in operation, both the State and the Federal governments need to insure orderly growth and to help minimize possible impact problems. The benefits of State and federal assistance to local communities may depend on the successful intergovernmental cooperation, including the participation of local governments and individual citizens.

On the positive side is the production of hydrocarbon fuels and the creation of many new job opportunities. From a negative perspective, however, the synfuels industry is labor-intensive, stimulating rapid population growth in rural areas and all the related socio-economic problems. Early government involvement in the development of such an industry is required, and the State of Colorado has led the way in bringing this issue to the attention of federal decision-makers.

AN UNCERTAIN FUTURE

The Federal government has leased only a small part of its extensive oil shale land holdings, and the amount and type of future leasing is still uncertain. Production operations are faced with a complex maze of regulations, permits and clearances involving considerable delays and duplication of paperwork at different levels of government. Capital and operating costs may be expected to be extremely high (one billion dollars or more) for most commercial facilities, and anticipated cost escalations make even these estimates uncertain.

At present the cost of producing a barrel of shale oil (including a fair return on investment) cannot compete with the production and investment costs of a barrel of conventional crude oil. Financing the costly facilities for oil shale development will prove difficult because of the many economic uncertainties: tax levies and their effects on an untested industry; environmental and community mitigation costs; and future costs of alternative energy fuels, to name but a few. The commercial development of an oil shale industry, indeed, will be a formidable task.

The future of oil shale development, like the mirages that float over the arid oil shale lands, seems only to recede as we approach it.

References: Oil Shale

Allred, V. D., 1967. Shale oil developments--kinetics of oil shale pyrolysis: Colorado School of Mines Quarterly, v. 62, no. 3, Golden, Colorado.

Banks, C. E., and Franciscotti, B. C., 1976. Resource appraisal and preliminary planning for surface mining of oil shale in Piceance Creek Basin, in The Ninth Oil Shale Symposium, Golden, Colorado, Proceedings: Colorado School of Mines, Golden, Colorado.

Cameron Engineers, 1978a. Shale oil status report, prepared for the Rocky Mountain Oil and Gas Association: Denver, Colorado.

Cameron Engineers, 1978b. Status of synfuels projects: Synthetic Fuels, Denver, Colorado, v. 15, no. 1.

Carver, H. E., 1964. Conversion of oil shale to refined products: Colorado School of Mines Quarterly, v. 59, no. 3, Golden, Colorado.

Chappell, W. R., 1978. Toxic trace elements and shale oil production, in The 11th Oil Shale Symposium, Golden, Colorado, Proceedings: Colorado School of Mines, Golden, Colorado.

Colorado Mining Association, 1977. State of Colorado mineral severance tax, in Mining Year Book, 1977, Denver, Colorado.

Colorado School of Mines Research Institute, 1975. A practical approach to development of a shale oil industry in the United States, prepared for Gary Operating Co.: Golden, Colorado.

Colorado Socio-Economic Impact Office, 1978. Colorado energy assistance plan: Denver, Colorado, January 1978.

Cook, C. W., 1974. Surface rehabilitation of land disturbances resulting from oil shale development: Executive Summary, Environmental Resources Center Information Series No. 11, Colorado State University, Fort Collins, Colorado.

Cook, W. C., 1970. Thermal analysis of oil shales: Colorado School of Mines Quarterly, v. 65, no. 4, Golden, Colorado.

Coomes, R. M., 1978. Carcinogenic aspects of oil shale, in The Environmental Aspects of Non-conventional Energy Resources-II, Proceedings, American Nuclear Society Topical Meeting, September 26-29, 1978; Denver, Colorado, pp. 26-1 -- 26-29.

Costello, J., 1978. Health studies of oil shale workers, in The 11th Oil Shale Symposium, Golden, Colorado, Proceedings: Colorado School of Mines, Golden, Colorado.

Davidson, D., 1976. The permit procedure--many steps, many stops: Shale Country, Mountain Empire Publishing, Inc., v. 2, no. 11, Denver, Colorado.

Davis, G. H., and Wood, Leonard A., 1974. Water demands for expanding energy development: U.S. Geological Survey Circular 703, Washington, D.C.

Energy Research and Development Administration, 1977. Environmental development and plan: Oil Shale - FY 1977, EDP/F-01(77), June 1977.

Environmental Protection Agency, 1974. Oil shale accomplishment plan, Region VIII, Denver, Colorado.

Environmental Protection Agency, 1977. Oil shale and the environment, EPA-600/9-77-033, October, Washington, D.C.

Environmental Protection Agency, 1978. EPA program status report: oil shale, EPA-600/7-78-020, February, Washington, D.C.

Ertl, Tell, 1965. Mining Colorado oil shale: Colorado School of Mines Quarterly, v. 60, no. 3, Golden, Colorado.

Federal Energy Administration, 1977. A report: regional profile of energy impacted communities, July 1977, Washington, D.C.

Ferris, B. J., 1948. Studies of soluble material in oil shale, in Mines Magazine, v. 38, no. 9, Golden, Colorado.

Grossman, A. P., 1977. Economic evaluation of combined in-situ and surface retorting of oil shale, in The Tenth Oil Shale Symposium, Golden, Colorado, Proceedings: Colorado School of Mines, J. H. Gary, Ed.

Hale, I., 1976. Energy sources '76: Denver Security Traders Assoc., Denver, Colorado.

Hebb, D. H., and Morse, J. G., 1976. Colorado energy resource handbook, vol. 1--Coal: Colorado School of Mines Publications, Golden, Colorado.

Hite, R. J., and Dyni, J. R., 1967. Potential resources of dawsonite and nahcolite in the Piceance Creek Basin: Colorado School of Mines Quarterly, v. 62, no. 3, Golden, Colorado.

Hopkins, J. M., Huffman, H. C., Kelley, A., and Pownall, J. R., 1976. Development of Union Oil Company upflow retorting technology, in The 81st National Meeting, April 11-14, Proceedings: American Institute of Chemical Engineers, Kansas City, Missouri.

Hoskins, W. J., Upadhyay, R. P., Bills, J. B., and Sandberg, C. R. III, 1976. A technical and economic study of candidate underground mining systems for deep, thick oil shale deposits, in The Ninth Oil Shale Symposium, Golden, Colorado, Proceedings: Colorado School of Mines Publications, Golden, Colorado.

Hubbard, A. B., and Robinson, W. E., 1950. A thermal decomposition study of Colorado oil shale: U.S. Bureau of Mines: Report of Investigations 4744, Washington, D.C.

Hutchins, J., 1978. Personal communication: September 20, 1978.

Jones, R. O., Mead, W. J., and Sovensen, P. E., 1978. Economic issues in oil shale leasing policy, in The 11th Oil Shale Symposium, Golden, Colorado, Proceedings: Colorado School of Mines Publications, Golden, Colorado.

Katell, S., and Wellman, P., 1974. An economic analysis of oil shale operations featuring gas consumption retorting: U.S. Bureau of Mines Technical Progress Report no. 81, Washington, D.C.

Kinney, P. W., 1977. Environmental permit directory: Office of Richard D. Lamm, Governor of Colorado, Denver, Colorado.

Ludlam, L. L., and Nutter, J. F., 1977. Perspective on modular demonstration of oil shale technology, in The Tenth Oil Shale Symposium, Golden, Colorado, Proceedings: Colorado School of Mines Publications, J. H. Gary, Ed.

National Petroleum Council, 1973. U.S. energy outlook: oil shale availability, Washington, D.C.

Novak, A., 1976. The shale paperwork maze of permits, clearances, regulations: Shale Country, Mountain Empire Publishing, Inc., v. 2, no. 11, Denver, Colorado.

Office of Technology Assessment, 1978. Draft working paper for a case study of oil shale retorting technology, Washington, D.C.

Paxton, J., 1976. Environmental studies: Shale Country, Mountain Empire Publishing, Inc., v. 2, no. 12, Denver, Colorado.

Project Independence, 1974. Project Independence report: Federal Energy Administration, November, Washington, D.C.

Project Independence Oil Shale Task Force, 1974. Potential future role of oil shale--prospects and constraints: U.S. Department of Interior, November, Washington, D.C.

Ridley, R. D., 1974. In-situ processing of oil shale: Colorado School of Mines Quarterly, v. 69, no. 2, Golden, Colorado.

Rio Blanco Oil Shale Project, 1976. Detailed development plan (Tract C-a): Gulf Oil Corporation and Standard Oil Co. (Indiana).

Ritz, W. R., 1978. Oil shale reserve may close: The Denver Post, September 20, p. 3, Denver, Colorado.

Rocky Mountain Oil and Gas Association, 1975. Summary of industry oil shale environmental studies and selected bibliography of oil shale environmental references: Denver, Colorado.

Shell Oil Co. and Ashland Oil, Inc., 1976. Oil shale tract C-b socio-economic assessment--impact analysis, v. 2, Detailed Development Plan, prepared for Area Oil Shale Office: Department of Interior, Grand Junction, Colorado.

Shih, N., 1978. Battelle Northwest (unpublished).

Sladek, T. A., 1974. Recent trends in oil shale, part 1: Mineral Industries Bulletin, Colorado School of Mines, v. 17, no. 6, Golden, Colorado.

Sladek, T. A., 1975. Recent trends in oil shale, part 2, Mining and shale oil extraction processes: Mineral Industries Bulletin, Colorado School of Mines, v. 18, no. 1, Golden, Colorado.

Sladek, T. A., 1978. Personal communication, September 28, 1978.

Spence, H. M., 1976. Severance taxes - what, why, how, when?: Shale Country, Mountain Empire Publishing, Inc., v. 2, no. 4, Denver, Colorado.

Stanfield, K. E., et al., 1951. Properties of Colorado oil shale: U.S. Bureau of Mines Report of Investigations 4825, Washington, D.C.

Strang, M. L., 1974. Committee on oil shale, coal, and related minerals--Report to the Governor and Colorado General Assembly: Legislative Council Publication 208, Denver, Colorado.

Stermole, F. J., 1974. Economic evaluation and investment decision methods: Investment Evaluations Corp., Golden, Colorado.

Sullivan, R. F., Strangeland, B. E., and Framkin, H. A., 1978. Refining shale oil, in The American Petroleum Institute's 43rd Midyear Meeting, Toronto, May 10: Washington, D.C.

Synfuels Interagency Task Force, 1975. Recommendations for a synthetic fuels commercialization program: technology and recommended incentive, v. III.

Tiedeman, J. A., and Terwilliger, C., Jr., 1978. Phyto-edaphic classification of the Piceance Basin: Colorado State University, Range Science Department, Ft. Collins, Colorado.

University of Oklahoma, 1975. Energy alternatives: a comparative analysis: Science and Public Policy Program, Norman, Oklahoma.

U.S. Department of Interior, 1978. Draft west-central Colorado coal environmental statement: Bureau of Land Management, Denver, Colorado.

Vranesh, G., and Cope, J. A., 1977. Water for oil shale development in western Colorado: a legal update - 1977, in The Tenth Oil Shale Symposium, Golden, Colorado, Proceedings: Colorado School of Mines, Golden, Colorado.

West, J., 1977. Obstacles to limit U.S. shale production: Oil and Gas Journal, August, v. 75, no. 35.

Appendix B: Oil Shale Leasing Program Tracts and Reserves

TABLE B.1

PROTOTYPE OIL SHALE PROGRAM LEASES IN COLORADO

Federal Tract	No. of Acres	Estimated Shale Oil Reserves in Barrels* Upper Zone	Lower Zone	Total
C-1	5120	614,400,000	4,505,600,000	5,120,000,000
C-2	5120	no shale above 30GPT	1,024,000,000	1,024,000,000
C-3	5120	307,200,000	1,024,000,000	1,331,200,000
C-a**	5090	610,800,000	4,479,200,000	5,090,000,000
C-6	5018	1,003,600,000	5,018,000,000	6,021,600,000
C-9	5128	1,180,000,000	None	1,180,000,000
C-10	5126	718,000,000	7,176,000,000	7,894,000,000
C-11	5118	1,535,000,000	7,677,000,000	9,212,000,000
C-12	5120	no shale above 30GPT in seams greater than 10 feet		
C-b***	5094	1,426,000,000	2,139,000,000	3,565,000,000
C-14	5120	1,178,000,000	None	1,178,000,000
C-15	5120	1,178,000,000	None	1,178,000,000
C-16	5120	921,600,000	5,376,000,000	6,298,000,000
Total Estimated Resource on all Federal Tracts in Colorado: (barrels of shale oil from oil shale grading at least 30 GPT)				49,091,800,000

*Shale grade of at least 30 gallons per ton (GPT)
**Includes Federal Tracts C-4, C-5, C-7, C-8, and C-17
***Federal Tract C-13

SOURCE: Project Independence Oil Shale Task Force, 1974, Appendix I.

TABLE B.2

RESERVE REQUIREMENTS TO SUPPORT OIL SHALE OPERATIONS

The following is the oil shale reserve base that is required to support the shale oil operations indicated for a period of 30 years.

Size of Operation (barrels per day)	Cumulative Reserve Base Required (barrels)
50,000	547,500,000
100,000	1,095,000,000
150,000	1,642,500,000
200,000	2,190,000,000
250,000	2,737,500,000

SOURCE: Project Independence Oil Shale Task Force, 1974, Appendix I.

Appendix C: Status of Commercial Oil Shale Projects as of May 1978

C-b SHALE OIL VENTURE -- Ashland (25%) & Occidental (75%)

Bonus bid of $117.8 million paid to acquire rights to tract in 1974. Original partners, ARCO and TOSCO, withdrew in 1975. A third original partner, Shell, withdrew 11/76. Occidental joined (with Ashland as remaining partner) 11/76. Modified DDP for 57,000 BPD modified in situ plant submitted March 1, 1977. DDP approved 8/30/77. EPA issued conditional PSD permit 12/16/77. Contracts have been awarded and work has begun. Primary contractor is Ralph M. Parsons Company. Collaring and cementing has been completed at 65' for the service shaft and 70' for the production shaft and the ventilation and escape shaft; actual shaft mining has begun.

Project Cost: $650 million

COLONY DEVELOPMENT OPERATION -- ARCO AND TOSCO

Proposed 47,000 BPD project on Colony Dow West property near Grand Valley, Colorado. Underground room-and-pillar mining and TOSCO II retorting planned. Production would be 66,000 TPD of 35 GPT shale from a 60-foot horizon in the Mahogany zone. Development suspended 10/4/74. Draft EIS covering plant, 196-mile pipeline to Lisbon, Utah, and minor land exchanges released 12/17/75. Final EIS has been issued. BLM conducting study on land exchange required for project. Study to be completed May 1978. World price for shale oil and inclusion of shale oil in entitlements program increases likelihood that project will be reactivated. If a proposed $3/bbl tax incentive becomes law, Colony hopes the climate will improve to attract enough investment for reactivation of the project.

Project Cost: Estimated at $1.132 billion including $20 million for community development.

OCCIDENTAL OIL SHALE, INC.

Oxy is developing their modified in situ retorting technology on their Logan Wash site near De Beque, Colorado. Field tests have been underway since 1972. Initial tests were conducted on three small retorts measuring 30 feet square by 70 feet high. Tests are now being conducted on commercial scale retorts measuring 120 feet by 280 feet high. Thirty thousand barrels of oil were produced from first commercial retort between December 75 and June 76. A $60.5 cost-sharing contract was signed 9/30/77 with DOE. Retort number 6 was rubblized 3/25/78. Production from retort 5 was 11,287 barrels.

Project Cost: To date at least $45 million spent
$60.5 million DOE cost-sharing contract

RIO BLANCO OIL SHALE COMPANY -- Gulf & Standard (Indiana)

Proposed project on Federal Tract C-a in Piceance Creek basin, Colorado. Bonus bid of $210.3 million to acquire rights to tract; lease issued 3/1/74. Revised DDP calling for use of LLL Rubblized in Situ Extraction of shale oil submitted to Interior 5/77. Combination of modified in situ retorts and TOSCO II surface retorts will be used to produce 76,000 BPD. Five year process development project will be conducted to prove in situ technology. Commercial facility scheduled to get underway in 1987. DDP approved 9/22/77. American Mine Services Inc. of Denver was awarded a $4 million contract 11/21/77 to sink a 15-foot wide, 976-foot deep shaft. EPA awarded PSD permit on 12/16/77. Contracts have been awarded and work has begun. Primary contractor is Morrison-Knudsen Company with a $38.8 million contract. Tests are underway to determine underground water quantities.

Project Cost: Four-year process development phase budgeted at $93 million
No cost estimate available for commercial facility

SUPERIOR OIL CO.

Proposed project involving production of shale oil, nahcolite, alumina and soda ash from a 6,500-acre privately owned tract in Piceance Creek basin near Meeker, Colorado. Underground mining and above-ground processing to yield shale oil, nahcolite, aluminum trihydrate, and soda ash. Facilities proposed to be constructed in modules of 13,300 BOPD from 28,000 TPD shale feed. Co-products would be 4,400 TPD of 80 percent nahcolite, 700 TPD alumina, and 1,500 TPD soda ash. Land exchange request to block up economically viable property filed with Interior 12/73. USGS appraisal of mineral values involved in request has been completed. BLM issued one-year contract to Environmental Research and Technology, Inc., in October 1977 for EIS. Work continues on the EIS in pursuit of the land exchange.

Project Cost: $300 million for one multi-mineral module
$473,459 for EIS

TOSCO SAND WASH PROJECT -- TOSCO Corp.

Proposed 75,000 BPD project on 14,688 acres of state leases in Sand Wash area of Uinta basin near Vernal, Utah. State-approved unitization of 29 non-contiguous leases requires $8 million tract evaluation by 1985. Minimum royalty of $5 per acre begins in 1984 and increases to $50 per acre in 1993. Preliminary feasibility and engineering work underway for modified in situ development in conjunction with surface retorting. Environmental assessment underway on site, but no other field work being conducted. TOSCO is planning to drill a core hole this fall on the Sand Wash site as a preliminary step to shaft sinking and establishment of a test mine. The test mine would confirm economics and mining feasibility plans for the commercial project.

Project Cost: Approximately $1 billion

UNION LONG RIDGE PROJECT -- Union Oil Company of California

In 1/74, Union announced plans for a commercial project ranging in size from 50,000 BPD to as much as 150,000 BPD on some 22,000 acres of fee land near Grand Valley, Colorado. Land, shale and water resources are adequate. Underground room-and-pillar mining and Union "B" retorting would be employed. Union's "B" retort is a modification of their direct-heated, rock pump retort first tested in the late 1950's. Current plans are to proceed with a 9,000 BPD (10,000 TPD) prototype facility before expanding to commercial production. Environmental and engineering studies are substantially completed for prototype facility. Union has announced that it will proceed if a $3/bbl tax credit is enacted.

Project Cost: Approximately $100 million for 9,000 BPD module

WHITE RIVER SHALE PROJECT -- Sun, Phillips & Sohio

Proposed joint development of federal lease tracts U-a and U-b in Uinta basin near Bonanza, Utah. Bonus bid for Tract U-a was $76.6 million by Sun and Phillips. Bonus bid for Tract U-b was $45.1 million by White River Shale Oil Corp. (jointly owned by Sun, Phillips and Sohio).

Rights to Tract U-b subsequently assigned to Sohio in exchange for portion of adjoining Skyline property and other considerations. Both leases issued 6/1/74. DDP filed with Interior 6/76 proposes modular development with ultimate expansion to 100,000 BPD. Room-and-pillar mines on both tracts. Common plant facility employing Paraho and TOSCO II retorting. Application for one-year suspension of operations granted 10/76. Injunction has been issued by U.S. Court in Salt Lake City preventing enforcement of terms of lease. Ownership of White River's land is clouded by unpatented pre-1920 oil shale placer claims, and by application for lease to some land by Peninsula Mining under Utah's in lieu land selections. A notice of appeal was filed on 8/29/77, but dropped on 4/11/78, leaving the preliminary injunction without contest. A final Environmental Baseline Report was issued by Interior on 11/15/77. Utah approved White River Dam and hydroelectric project 2/78.

Project Cost: Estimated at $1.61 billion for 100,000 BPD project

SOURCE: Cameron Engineers, 1978b, pp. B-1,2.

Part III

Uranium

19
Nature of the Resource

Uranium is the heaviest naturally-occurring element. A piece of uranium the size of a 12-ounce soft drink weighs about 17 pounds, about 65 percent more than lead. It is a very dense, highly chemical-reactive metal that is also radioactive. First isolated as the dioxide by Klaproth in Germany in 1789, it was named for the planet Uranus (Singleton, 1968). Uranium was used primarily in high temperature dyes until Hahn and Strassman, in 1939, observed that it underwent fission when exposed to neutrons of discrete energy levels. From relative obscurity, the element suddenly vaulted to center stage. Shortly thereafter, Fermi and others in the United States demonstrated that under proper conditions the fission process could become self-sustaining; that is, a nuclear chain reaction resulted, capable of the controlled release of large quantities of energy. For a more detailed discussion of atoms, radioactivity, and the fission process, see Appendix A.

Naturally-occurring uranium is made up of three isotopes (atoms with the same chemical characteristics but of different atomic mass). These are: U-238, 99.283 percent of the uranium found in nature; U-235, 0.711 percent; and U-234, the remaining 0.006 percent. Thus, in 100,000 uranium atoms, approximately 99,283 are U-238, 711 are U-235, and 6 are U-234.

The only naturally-occurring, readily fissionable isotope is U-235. Theoretically, the heat energy released by complete fission of one pound (454 g) of U-235 is equivalent to the energy released by burning approximately 2.5 million pounds of coal. In practice, however, since the thermal efficiency of a nuclear power plant is usually about 33 percent, it requires at least 3 - 4 pounds of U-235 to produce the equivalent energy in the form of electricity (Committee on Resources and Man, 1969, p. 219).

Two man-made isotopes, plutonium-239 (Pu-239) and uranium-233 (U-233), can also provide energy through the fission process in a nuclear reactor. (Note: Neither are found in nature.) These isotopes are created from the "fertile" materials U-238 and thorium-232, respectively. (A fertile material is not fissionable, but it can be converted (transmuted) to a fissionable material by absorbing neutrons such as found in a nuclear reactor, particularly in a breeder reactor.) However, neither thorium nor U-238 can be converted easily nor rapidly to the fissionable isotopes U-233 or Pu-239 in present power reactors; most existing reactors produce only small quantities of these two materials. Consequently for many years to come, we will probably have to depend upon the small fraction of U-235 in natural uranium to provide our nuclear fuel needs. (For a more detailed discussion, see chapter "Technology Developments Affecting Uranium Use.")

The principal needs for uranium are for use in civilian nuclear power plants and for military requirements. However, uranium is also an important raw material in the production of various radio-isotopes used in medicine, chemistry, and industry. Uranium is also used in minor amounts in the following applications: as a "getter," to purify hydrogen and inert gases; in a variety of electrical and electronic devices; and as an alloy additive to improve corrosion resistance and high-temperature properties of steel (Griffith, 1967, pp. 3-5). Finally, uranium is still used as a coloring agent for glass and ceramic glazes, as it was before 1940 (U.S. Geological Survey, 1974, p. 7).

URANIUM MINERALS

Uranium, although one of the less common elements, is widely distributed and constitutes about two to three parts per million of the earth's crust. It is more abundant than gold, silver, or mercury; as abundant as tin; and less abundant than copper, lead, and zinc (Nininger, 1954, p. 15). Like most other metallic elements, uranium does not occur in nature as a free metal. Since it is highly reactive chemically, it is found combined with oxygen or various other elements and forms both soluble and insoluble compounds. The soluble compounds are readily transported and become diffused throughout the earth's crust by surface and ground waters. Consequently, uranium is widely distributed throughout the world, occurring in a wide range of rocks and mineral deposits. However, although uranium is widespread, it is not often found in large orebodies that are as rich as those formed by other metals of similar abundance. In fact, most large uranium deposits have a uranium content of only about 0.1 percent (Griffith, 1967, p. 1).

Nearly 100 different minerals are known to contain uranium. Its mineral forms can be broadly classified as primary and secondary. Primary uranium minerals were deposited at the same time as the rocks in which they occur; secondary uranium minerals were formed later by the dissolving and redeposition of chemically-altered primary minerals. The most important primary minerals are uraninite (UO_2, also called pitchblende) and coffinite ($U(SiO_4)_{1-x}(OH)_{4x}$). Nearly 95 percent of the uranium mined in the United States is found in pitchblende and coffinite deposits. Secondary uranium minerals are numerous and diverse, including hydrated oxides, phosphates, arsenates, vanadates, sulfates, carbonates, and silicates. The most common and commercially valuable forms of secondary uranium minerals are carnotite and tyuyamunite, each containing vanadium. Most secondary uranium minerals are brightly colored (usually yellow, orange, or green), and can be readily identified. In fact, Indians used carnotite ores as pigments for war paint (Ballard and Conklin, 1955, p. 5).

There are three important types of uranium deposits:

1. "Sandstone type" - Here uranium ore is confined to a particular sandstone bed. These deposits were created millions of years ago when decaying plant material altered the subsurface environment, causing uranium dissolved in stream waters to precipitate. Since much of the original plant material has turned to mineral or stone, such "petrified rivers" often contain ore bodies of pitchblende, coffinite, carnotite, and other uranium minerals. Sandstone uranium deposits range in size from a few feet in width, containing a few tons of ore, to several thousand feet, with several million tons of ore. Average grade usually ranges from 0.1 to 0.3 percent uranium oxide (U_3O_8). Sandstone ores have been the principal source of uranium in the United States, but are less important elsewhere.

2. "Vein deposit" - A vein is a fissure (crack) in a rock formation, usually vertical or steeply inclined, containing minerals originating in molten solutions that were forced into fissures under great pressure.* Individual veins range from a few inches to several feet wide, but only a few deposits contain large quantities of uranium ore. Average uranium grade usually ranges from 0.1 to 1.0 percent.

3. "Ancient conglomerate" - This is the most important source of uranium outside the United States. Uranium minerals are concentrated as placer deposits in ancient quartz-pebble conglomerate beds. Although such deposits are usually large and wide-spread, average grade only ranges from 0.03 to 0.1 percent uranium. In the Witwatersrand area of South Africa, conglomerate ores contain only 0.03 to 0.07 percent uranium, but uranium is still recoverable as a by-product of gold mining (U.S. Geological Survey, 1974, p. 10).

AMERICA'S URANIUM DEPOSITS

The principal higher-grade uranium deposits in the United States lie in three regions: the Colorado Plateau, the Wyoming

---

*This is the major theory of uranium vein deposits. Another theory is that sulfides caused deposition of uranium in ground water percolating through fractures in the veins (D. B. Collins, 1977).

Basins, and the Gulf Coastal Plain (See Figure 19.1). There are many other uranium mining areas and mineral belts, both inside and outside these regions; however as shown in Table 19.1, just two of these regions--the Colorado Plateau and the Wyoming Basins--accounted for 91 percent of the total United States production of U_3O_8 (uranium ore concentrate, also called yellow-cake) in 1977.

TABLE 19.1

DISTRIBUTION OF 1977 U_3O_8 PRODUCTION IN ORE BY RESOURCE REGION

Resource Region	Tons U_3O_8	% Total Tons U_3O_8
Colorado Plateau	9,900	59
Wyoming Basins	5,300	32
Others: Northern Rockies, Southern Basin and Range, W. Gulf Coastal Plain, Colorado & Southern Rockies	1,500	9
Total	16,700	100

SOURCE: U.S. Department of Energy, 1978a, p. 15.

Note: Includes miscellaneous receipts.

There are also several low-grade uranium deposits in the United States. Beds of uranium-bearing marine phosphorite, commonly five to ten feet thick, underlie hundreds of square miles of land in Idaho, Montana, Utah, Wyoming, and Florida (U.S. Geological Survey, 1974, p. 8). Chattanooga Shale is another low-grade source of uranium; beds approximately fifteen feet thick underlie a sizable fraction of the area made up by Tennessee, Kentucky, Ohio, Indiana, and Alabama. Both of these deposits contain millions of tons of uranium. The average grade in this area is 0.006 percent--somewhere between 1/20th and 1/50th as rich as the average high-grade ore (-.1 to 0.3 percent) found in the western states (Battelle, 1974, p. 6.54). Thus it probably would not pay to work such low-grade deposits for quite some time to come.

A National Uranium Resource Evaluation (NURE) program, begun by the United States Atomic Energy Commission (AEC) in 1974, is being continued by the United States Energy Research and Development Agency (ERDA), now part of the United States Department of Energy (DOE). The NURE program's goal is to evaluate uranium resources and to identify areas favorable for uranium exploration. Figure 19.2 shows twenty designated

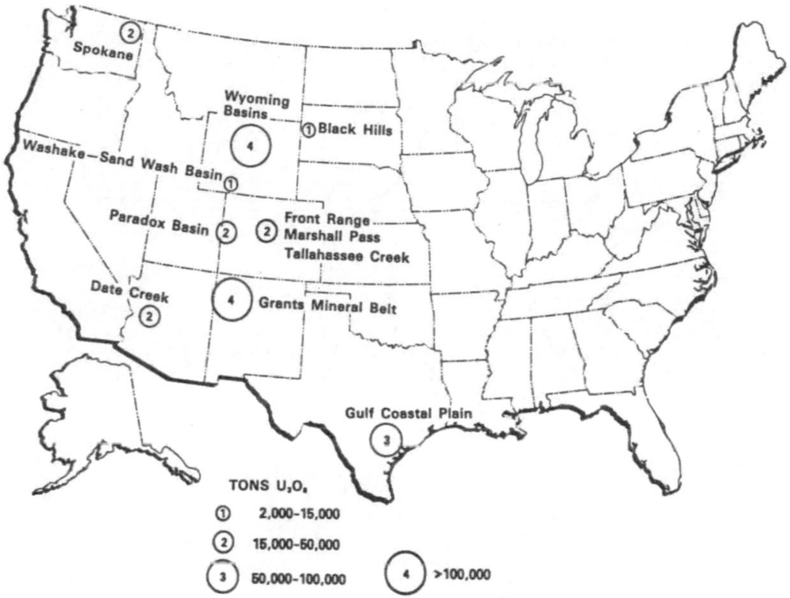

Figure 19.1 Distribution of January 1, 1978 $50 Reserves (major districts)
Source: Meehan, 1978.

Figure 19.2 Uranium Resource Regions of the United States
Source: ERDA, 1977a, p. 20.

uranium resource regions, based on geologic considerations.

DOE differentiates between ore reserves and potential resources in reporting uranium resources. Ore reserves, marked by drilling or other direct sampling methods, make up the "best known" or "most certain" portions of the uranium resource. Ore reserves are defined as "that quantity of ore, in identified deposits, that can be developed at the current levels of technology and current costs and prices."

Potential resources, on the other hand, are judgment estimates by geologists of the quantity of uranium believed to be present in deposits that are either incompletely known or undiscovered. There are three classes of potential resources: probable, possible, and speculative, in order of declining reliability. Ore reserve calculations are reliable within a range of plus or minus 20 percent; potential resource calculations are less reliable, but just how much less, has not been fixed (ERDA, 1976a, p. 10). For a detailed definition of probable, possible, and speculative potential resources, please consult the glossary of this report. Figure 19.3 shows the relationship between ore reserves and potential resources.

NURE estimates of both ore reserves and potential resources are based upon "forward costs," that is, those operating and capital costs which would be incurred in developing a uranium property at the time the estimate is made. As such, they include "first destination" (mine-to-mill) production costs (labor, materials, ore transport, power, royalties, taxes other than income, mine development, and mill construction). They do not include a profit factor; nor do they include costs already incurred, such as property acquisition, exploration, or previous mine development. Consequently, the actual total production cost to a company, including recovering spent capital and profit-making, can considerably exceed the stated "forward cost." In fact, a study done in 1974 found that ". . . it is easy to assume conditions under which so-called '\$8/pound U_3O_8' (as defined by forward cost method) must be sold at \$13 - \$14 per pound to yield a fair after-tax profit to the producer" (Battelle, 1974). (For a more complete discussion, refer to the section entitled "Uranium Price Considerations.") A summary of the potential uranium resource for the United States, based on "forward costs" of \$15, \$30, and \$50 per pound of U_3O_8 is shown in Table 19.2. Note that as the cost increases (i.e., the number of dollars a producer is willing to pay to extract uranium), the quantity of the resource available also increases. Table 19.3 lists the estimated U.S. uranium mineral inventory by grade of ore. However, these estimates do not represent ore reserves, since much of the higher grade material is not presently economically recoverable due to depth, small deposit size, etc. Table 19.4 indicates the uranium ore reserves and potential resources, by resource region, for a \$30 forward cost. Note that only three of these regions (the Colorado Plateau, Wyoming Basins, and the Coastal Plain) contain 93 percent of the ore reserves and 68 percent of the potential uranium

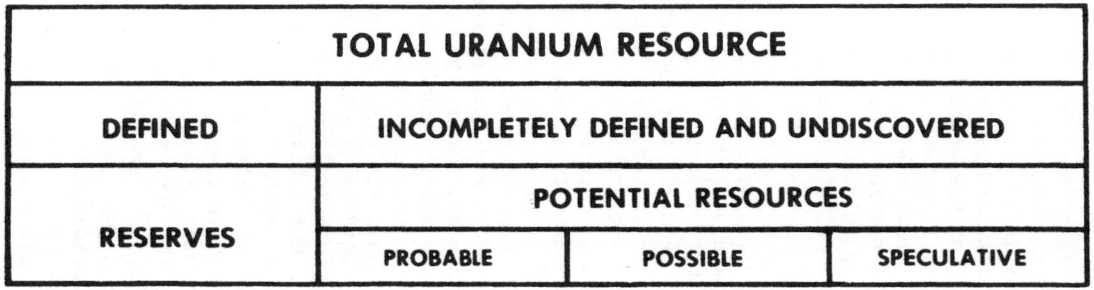

Figure 19.3 Uranium Resource Classifications
Source: U.S. ERDA, 1976 a, p. 9.

TABLE 19.2

URANIUM RESOURCES AS OF JANUARY 1, 1978

	Tons U_3O_8			
		Potential		
Cost Category	Reserves	Probable	Possible	Speculative
$15	370,000	540,000	490,000	165,000
$15-30 Increment	320,000	475,000	645,000	250,000
$30	690,000	1,015,000	1,135,000	415,000
$30-50 Increment	200,000	380,000	380,000	150,000
$50	890,000	1,395,000	1,515,000	565,000

SOURCE: Hetland & Grundy, 1978, p. 3.

Note: Uranium that could be produced as a byproduct of phosphate and copper production is estimated at 140,000 tons U_3O_8.

TABLE 19.3

UNITED STATES POSTPRODUCTION URANIUM MINERAL INVENTORY, 1/1/78

Mimimum Grade (% U_3O_8)	Cumulative Tons of Ore (Millions)	Avg. Grade (% U_3O_8) of Cumulative Tons	Cumulative Tons U_3O_8 (Thousands)
0.01	2,982	0.04	1,287
0.02	1,922	0.06	1,311
0.03	1,284	0.08	975
0.04	924	0.09	860
0.05	693	0.11	760
0.06	532	0.13	674
0.07	416	0.14	601
0.08	328	0.16	536
0.09	261	0.18	481
0.10	208	0.21	432
0.11	182	0.22	406
0.12	161	0.24	383
0.13	143	0.25	362
0.14	128	0.27	343
0.15	115	0.28	325
0.16	104	0.30	309
0.17	94	0.31	294
0.18	86	0.33	281
0.19	79	0.34	268
0.20	72	0.36	257
0.21	67	0.37	245
0.22	62	0.38	235
0.23	57	0.40	226
0.24	53	0.41	217
0.25-Over	49	0.42	208

SOURCE: U.S. Department of Energy, 1978a, p. 19.

Note: These figures do not represent ore reserves, since the economics of exploitation are not taken into account.

TABLE 19.4

SUMMARY OF URANIUM PRODUCTION, RESERVES, AND POTENTIAL BY REGIONS

Region	Tons U_3O_8 Production to 1/1/78	Tons U_3O_8 (\$50/lb) 1/1/78 Reserves	1/1/78 Potential Resources Probable	Possible	Speculative
Colorado Plateau	216,300	485,200	665,000	815,000	40,000
Wyoming Basins	68,900	264,100	375,000	115,000	30,000
Coastal Plain	10,000	53,900	180,000	95,000	35,000
Northern Rockies		25,400	27,000	63,000	50,000
Colorado and Southern Rockies	17,100	25,800	56,000	56,000	41,000
Great Plains		8,000	27,000	70,000	48,000
Basin and Range		25,500	59,000	292,000	76,000
Pacific Coast and Sierra Nevada	<1,000	2,100	4,000	9,000	9,000
Central Lowlands	<1,000	0	†	†	110,000
Appalachian Highlands	<1,000	0	†	†	95,000
Columbia Plateaus	<1,000	0	†	†	31,000
Southern Canadian Shield	0	0	†	†	†
Alaska	<1,000	0	2,000	†	†
TOTAL	313,100	890,000	1,395,000	1,515,000	565,000

<u>SOURCE</u>: Hetland & Grundy, 1978, p. 6.

† Resources not estimated because of inadequate knowledge.

resources. Consequently, for some time to come, these regions will probably continue to be where most United States production, development, and exploration for uranium takes place.

COLORADO'S URANIUM DEPOSITS

Colorado's uranium deposits occur in two geologically-distinct areas: western Colorado (west of the Continental Divide) and the Front Range (east of the Continental Divide). These two areas, along with the principal uranium districts, are shown in Figure 19.4.

Western Colorado is the oldest uranium mining area in the United States. Most deposits occur in sandstone formations, with uraninite, coffinite, and carnotite ores predominating. Uranium ore was first discovered in the Rock Creek area of Montrose County in 1881. The first uranium production in Colorado (ten tons of carnotite ore from the same areas) was shipped to the French scientists Marie and Pierre Curie in 1898 (U.S. Bureau of Mines, 1955, p. 8). This was only the beginning. From 1948 through 1974, western Colorado produced 12 percent of the total United States uranium ore mined. This amounted to 35,556 short tons of U_3O_8 from 15,589,100 tons of ore averaging 0.23 percent U_3O_8. In addition, 175,261 tons of vanadium oxide (V_2O_5) were recovered during the same period (Chenoweth, 1977, p. 1). Vanadium, used in steel alloys, does not always occur with uranium. When it does, however, as in the carnotite ores found in the Uraven Mineral Belt, it is usually recovered along with the uranium. While usually considered a by-product rather than a co-product, in the last several years the dollar value of the vanadium produced has been no less than 50 percent, and as high as 95 percent, of the value of the "primary" product uranium. Thus, it is likely that, for many of the Uraven Mineral Belt mining operations, the presence of vanadium enabled the extraction of uranium ores which otherwise would have been uneconomic to develop. Unfortunately, little or no vanadium is found in the uranium deposits in Colorado east of the Uraven Mineral Belt. Consequently, existing or potential uranium mines in the Cochetopa, Marshall Pass, and Maybell districts will have to rely solely upon the uranium content of a deposit to operate successfully.

In the Front Range, uranium deposits usually occur as veins, with small bodies of pitchblende commonly associated with sulfide minerals. Usually the uranium minerals are only a small fraction of the vein; the veins are chiefly valued for their gold content. As a result, with few exceptions there has been little activity connected with uranium in the Front Range area.

The Schwartzwalder Mine, located in Jefferson County, is a notable exception. It is the largest uranium mine in Colorado, producing approximately one-sixth of the State's uranium output (based on dollar value) in 1976. Another recent exception occurs in the Tallahassee Creek area where the Cyprus

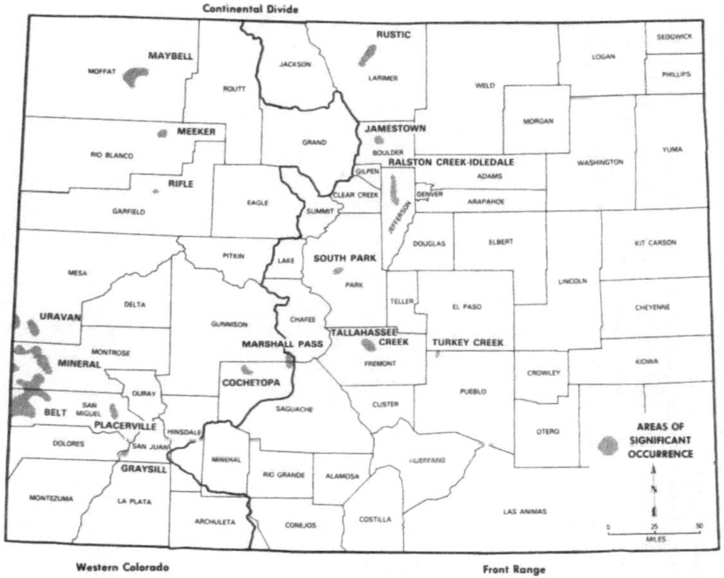

Figure 19.4 Uranium Districts in Colorado
Source: Chenowith, 1977 a, p. 2.

Mines Corporation announced discovery of a major uranium deposit, estimated to contain between five and ten thousand tons of U_3O_8 (The Mining Record, 1977, p. 1). On the whole, because of limited exploration, less is known about uranium ore reserves in the Front Range area than in western Colorado.

Table 19.5 gives potential uranium resources (DOE estimate) for Colorado. An estimate of the eastern Colorado resource is included to provide information about Front Range potential. Note: Total potential resource for the Front Range approximately equals that of western Colorado, but a greater proportion of Front Range potential occurs in the less reliable categories. Figure 19.5 shows areas in Colorado to which these potential resource estimates apply.

The uranium fuel cycle, as shown in Figure 19.6, includes all of the steps required for converting the naturally-occurring raw material into reactor fuel through storage of spent reactor wastes. The cycle begins with exploration for the fuel source material.

TABLE 19.5

COLORADO URANIUM RESOURCES (tons U_3O_8)

Area	Reserves	Potential Resource Probable	Potential Resource Possible	Speculative	Total Potential Resource
State of Colorado	NA	130,000	181,000	72,000	383,000*
Western Colorado	16,769	75,000	23,000	26,000	124,000
Front Range	NA	26,000	59,000	11,000	96,000

SOURCE: Hetland & Grundy, 1978; Chenoweth, 1977b.

*Forward cost of \$30/lb U_3O_8

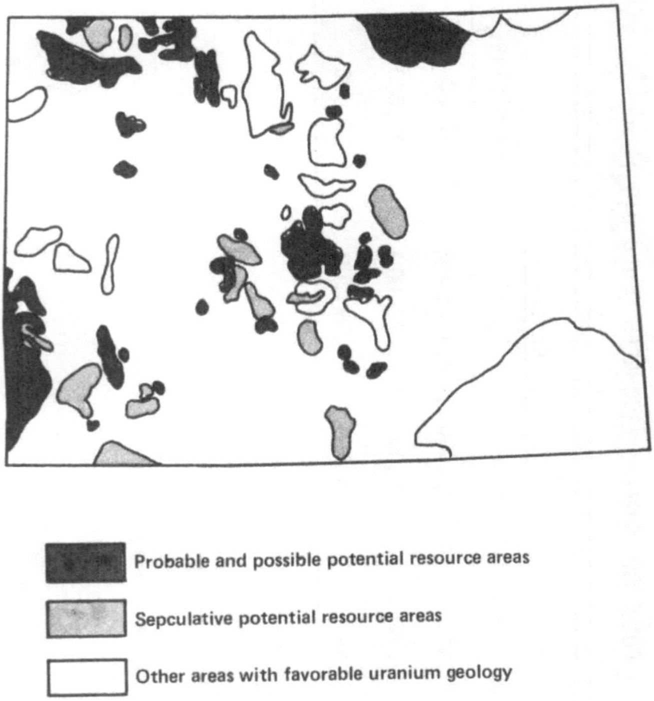

Figure 19.5 Potential Resource Estimates in Colorado
Source: Malan, 1977.

The Uranium Fuel Cycle

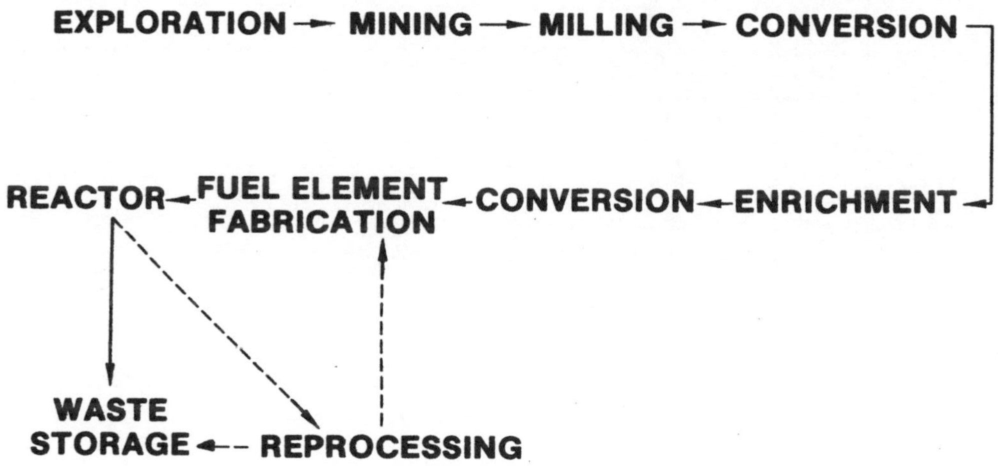

Figure 19.6

20 Exploration

Exploration includes the search for minerals, or indications of their possible existence, as well as the work necessary to determine if an economic ore body exists. In general, most uranium exploration programs are conducted in three phases: (1) preliminary regional reconnaissance to identify potential areas; (2) detailed reconnaissance to confirm and rank the favorable areas; and (3) intensive geophysical and geochemical exploration of selected favorable areas by drilling, sampling, etc. Approximate areas associated with each phase are: regional--up to tens of thousands of square miles; detailed reconnaissance--up to a few hundred square miles; intensive--several claims or just a few square miles. Early exploration methods designed to "narrow the field" were crude, consisting mainly of visual examination of surface outcrops and float material. Characteristic geological formations and oxidized ore colors were used as guides to determine the presence of uranium minerals. Subsequently, more rapid and efficient methods were developed to detect the presence of uranium.

TRADITIONAL METHODS

Since the 1940s, the search for uranium has been based on two characteristics: its geochemical activity and its radioactivity. Table 20.1 summarizes the traditional uranium exploration methods, based primarily on the latter property.

Uranium has a relatively high level of chemical activity, interacting with the subsurface environment, carbonated or acidic groundwater and hydrothermal solutions. Readily oxidized, uranium exists in one of several oxidation states, each with a known degree of solubility. In other words, the local acidity level will affect its oxidation state, and thus its solubility or insolubility. Consequently, under one set of conditions, uranium salts will dissolve and can migrate in subsurface waters; under another set of conditions, the uranium will precipitate and concentrate in a given area. As a result, water samples may often indicate the presence of uranium in unexposed deposits. In fact, one element of the National Uranium Resource Evaluation (NURE) program is the sampling of surface waters, stream sediments

TABLE 20.1

TRADITIONAL METHODS IN URANIUM EXPLORATION

1. Geological mapping	Use of aerial and satellite photography to select areas of high success probability for exploration.
2. Airborne radiometric surveys	Total gamma counting and spectrometric data (see text).
3. Ground surveys	Geological and radiometric measurements to locate suitable host rocks.
4. Radon measurements in soil, sub-soil and water	Anomalous concentrations of radon gas and its radioactive decay products can provide data on concentrations of uranium in porous soils. Measurement in streams, lakes and springs is also useful if knowledge exists of groundwater hydrology.
5. Geochemical surveys	Sampling of soils and stream or lake sediments for multi-element data can provide useful clues to the local geochemistry.
6. Geobotanical surveys	Although limited to suitable vegetation, the method can be useful to determine selective plant uptake or uranium, or associated minerals (e.g., selenium).
7. Other geophysical methods	Such parameters as lithology, structure and associated minerals are indirect indicators and may be useful.
8. Exploration drilling	Used when geological indications are favorable and drill holes are used for logging.

SOURCE: Bowie and Cameron, 1976, p. 5.

and ground waters to determine variations in uranium and other selected elements, as guides in the search for uranium.

The second characteristic of uranium, providing both a means of detection and a measurement of the amount present assuming equilibrium, is its radioactivity. Uranium's two principal isotopes are U-238 (99.3 percent) and U-235 (0.7 percent): each undergoes decay in its own fixed pattern, forming radioactive daughters whose ultimate fate is the formation of nonradioactive elements. Radioactive bismuth (Bi-214), formed in the U-238 decay series, shown in Figure 20.1, serves as the most readily detectable radio-element in the series because of the type of radiation (gamma), energy level, and abundance. Bi-214 normally indicates the presence of uranium, provided equilibrium conditions are maintained, i.e., uranium and its daughters remain immobilized close to each other.

Other radio-elements occurring naturally in the earth's crust are thorium (Th-232) and potassium (K-40). The presence of radiation in soils, often indicates the co-existence of U, Th, and K, each emitting radiation of specific types and at discrete energy levels. Using a technique called spectroscopy, one can selectively measure and distinguish each of the three species present by energy level discrimination. This method has been used in field exploration for more than a dozen years (Dodd and Eschliman, 1972; Wollenberg, 1977). In recent years, compact, easily portable instruments have become commercially available, and a down-hole probe is still in the development stage.

Meaningful field measurements are difficult to obtain because uranium always occurs as a compound, with the solubility of each compound varying, depending on the subsurface acidity and oxidation potential (pH and Eh). Because of this variable solubility, when detection instruments indicate high levels of radioactivity derived from uranium, the sought-after material may not be directly in view of the detector or even very close to it. What is being seen by the instrument is the most detectable daughter decay product; the uranium, in solution, may have been carried some distance away. For example, a high level of Bi-214 radioactivity does not always indicate the presence of a uranium ore body near the measured radioactivity: the ore may have moved elsewhere* (Saum and Link, 1969). To eliminate possible and often serious errors, prudent exploration calls for sending soil and rock samples to laboratories for specific chemical analyses. This procedure, although costly and time-consuming, is necessary, particularly when men and material are sent into the field. This analysis procedure normally accompanies the local surveys listed in Table 20.1 as part of standard exploration methods.

Instruments currently used in the field to detect the gamma rays emitted by uranium ore are either Geiger-Muller Counters

*"Roll front" formation (see Glossary).

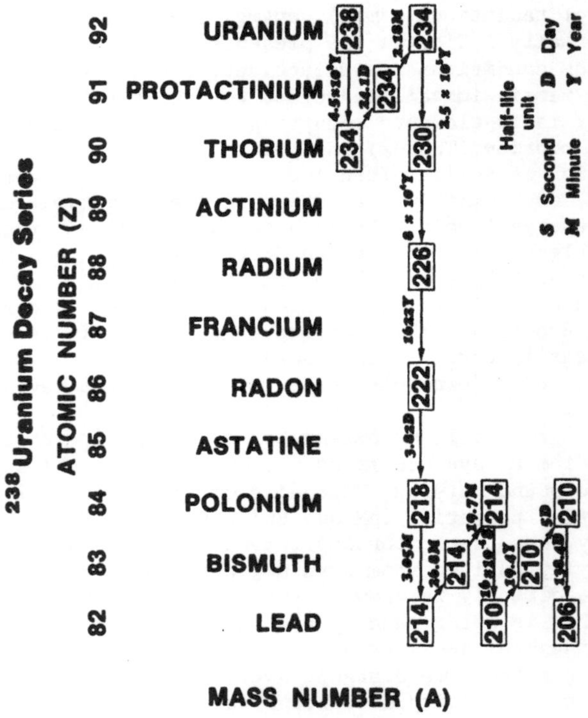
238 Uranium Decay Series
ATOMIC NUMBER (Z)
92 91 90 89 88 87 86 85 84 83 82
URANIUM
PROTACTINIUM
THORIUM
ACTINIUM
RADIUM
FRANCIUM
RADON
ASTATINE
POLONIUM
BISMUTH
LEAD
MASS NUMBER (A)
Half-life unit
S Second
M Minute
D Day
Y Year

Figure 20.1

(Geiger Counters) or scintillometers. Although both are sensitive to gamma rays, the scintillometer is more effective and more frequently used. Sophisticated circuitry in a scintillometer can measure the individual energy levels of gamma ray emissions--a basis for differentiating uranium, thorium, and potassium. However, radiometric prospecting is most effective in locating uranium deposits that are old and close to the surface. Where the deposit is recent (less than 500,000 years old) or where there is a thick overburden, this technique is less reliable.

ADVANCED METHODS

Unfortunately, most domestic large uranium deposits, readily detectable at the surface by traditional measurements of radioactivity, have already been discovered. Future exploration will have to use the more sophisticated detection methods. (See Table 20.2). Such advanced methods could locate more deeply buried ore bodies, whose surface radiations are tenuous at best. Locating these deposits would depend upon the more sensitive instruments coupled with computers to analyze the data obtained (Clayton et al., 1976). The exploration techniques used in the near future will probably be the product of integrating newer methods with traditional, well-established techniques, since no single method is likely to yield the "best" results.

EXPLORATION ACTIVITY

High prices and a continuing tight supply have greatly accelerated uranium exploration activities in the United States. Surface drilling, composed of exploration and development drilling totaled 41.0 million feet in the United States in 1977. Compared to 34.8 million feet in 1976, this is an increase of 21 percent (U.S. Department of Energy, 1978a, p. 55). This increase, however, appears to be taking place primarily in historically-productive districts or in similar sandstone environments nearby. Table 20.3 shows the percent by state distribution of surface drilling for uranium. Little mineralization of economic significance has been discovered recently outside familiar areas. The main reasons for this lack of success are probably inadequate funding, obsession with sandstone concepts, and poor understanding of uranium occurrences in other environments (Davis, 1977). Further, while exploration in "traditional" districts has been moderately successful, drill holes are getting progressively deeper (Figure 20.2).

Finally, exploration is not without its problems. One problem is the lack of experienced uranium geologists, partly a result of the "soft" uranium market in the late 1960s and early 1970s. Delays in securing environmental permits, costly in both time and dollars, are also hindering exploration efforts. While some difficulties have been encountered in acquiring land for exploration and production, they have not proved insurmountable, as indicated by the increasing land holdings shown in Table 20.4.

TABLE 20.2

NEW METHODS IN URANIUM EXPLORATION

Reconnaissance	
1. Radiogenic heat	Self absorption of radiogenic heat (emitted radiation) in near-surface uranium deposits leads to small increases in earth surface temperatures which may be measurable prior to earth heating by solar radiation (before dawn). Infrared scanning devices sensing $\pm$ 0.25^{o}C can be flown on aircraft to make these measurements. This method requires further development (1, 2).
2. Helium measurement	Uranium and several of its radioactive daughters decay by emitting alpha particles, with the latter converting to helium atoms. Surface mass spectrometric measurements over porous rock structures appear to provide a useful detection tool (3).
3. Remote sensing (satellite imagery)	Sophisticated computer analyses, using digital and analog enhancement of multispectral imagery, are being developed for unique identification of a few minerals found in association with uranium deposits. LANDSAT satellite imagery has been analyzed for discrete spectral signatures in 50 known uranium districts in western United States, to develop a further basis for exploration (4).

Direct Uranium Measurements - To eliminate serious inaccuracies due to the migration of uranium away from its highly radioactive daughter products, the measurement of uranium directly is being approached from several directions. The most promising appear to be:

TABLE 20.2 (continued)

1. In-place nuclear fission	A source of neutrons at the proper energy levels will induce nuclear fission in the U-235 isotope. Fission of uranium will result in the spontaneous emission of two classes of neutrons: prompt and delayed; both are readily detectable with existing instrumentation. Several neutron sources (radioisotopes and small electronic accelerators) are under t st to develop units for use in the field (5).
2. Gamma spectrometry	Present systems differentiate among radioactive daughters of U, thorium, and potassium. A more elegant approach is directed towards the measurement of lower energy emissions from U-daughters other than Bi-214 to more effectively relate U proximity to the detected radiations (6,7).
3. X-ray fluorescence	Useful in the drilling phase for delineation of the boundaries of a "roll-front" deposit by identifying "pointer" elements in the sandstone matrix. When combined with X-ray diffraction, these tools supply systematic mineral and elemental data to define the discovery (8).
4. Non-nuclear geophysical methods	Also under development (improvement), are indirect conventional geophysical methods including: electrical resistivity (SP), induced polarization (IP), gravity, seismic, electromagnetic pulsing (EM), conventional magnetic intensity and in-situ magnetic susceptibility (7).

<u>SOURCES</u>: (Derived from:)

1. Bowie and Cameron, 1976, p. 11.
2. Grutt, 1977, p. 66.
3. Grutt, 1977, p. 65.
4. Kober and Procter-Gregg, 1977.
5. Smith, 1977.
6. Senftle, et al., 1976.
7. Dodd, 1977.
8. Slaughter, 1977.

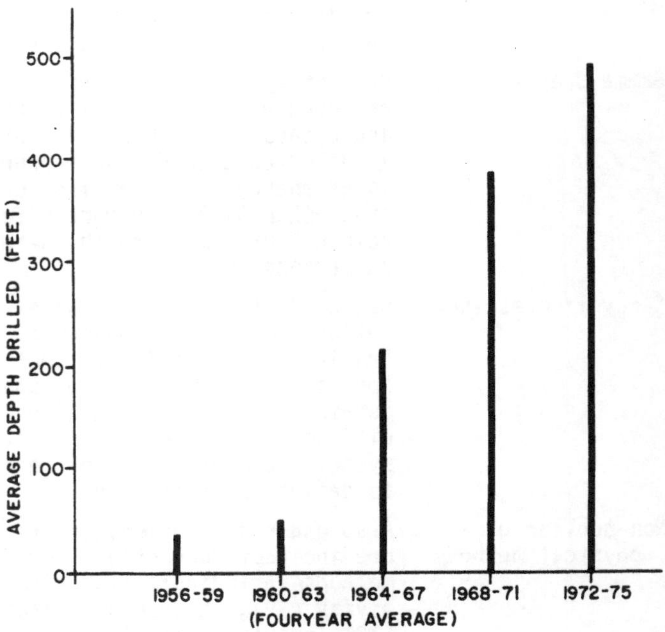

Figure 20.2 Exploration - Average Depth of Drillholes
Source: Davis, 1977.

TABLE 20.3

DISTRIBUTION OF 1977 DRILLING BY STATE (Millions of Feet)

State	Drilling	% of Total
Wyoming	15.4	37.6
New Mexico	9.1	22.2
Texas	5.8	14.2
Utah	3.7	9.0
Colorado	4.0	9.7
Others*	3.0	7.3
Total	41.0	100.0

SOURCE: U.S. Department of Energy, 1978a, p. 56.

*Includes Alaska, Arizona, California, Idaho, Michigan, Montana, Nevada, North Carolina, North Dakota, Oklahoma, Oregon, Pennsylvania, South Dakota, and Washington.

In summary, if the increased demand and the price of uranium continues, exploration efforts, both in terms of dollars and feet drilled, should continue to increase. Such exploration will probably be characterized by using advanced, as well as traditional, exploration techniques and investigations into less familiar areas.

TABLE 20.4

LAND HELD FOR URANIUM EXPLORATION AND MINING (in thousands of acres)

Distribution by State					
	1/1/74	1/1/75	1/1/76	1/1/77	1/1/78
Arizona	754	819	942	1,021	1,212
California	587	619	619	631	631
Colorado	1,291	1,592	1,623	1,852	2,431
Idaho	34	70	81	108	138
Montana	380	438	418	420	488
Nevada	264	312	321	376	478
New Mexico	3,158	3,378	3,663	3,885	3,855
North Dakota	100	100	100	128	256
Oregon	31	31	31	31	31
South Dakota	81	91	87	810	882
Texas	641	627	622	676	798
Utah	2,783	3,515	4,185	5,498	5,829
Washington	72	76	129	401	414
Wyoming	8.598	9,608	10,090	11,246	11,868
Total	18,774	21,276	22,911	27,083	29,311

SOURCE: U.S. Department of Energy, 1978a, p. 68.

21 Uranium Recovery

CONVENTIONAL MINING METHODS

Following exploration, during which a uranium deposit is outlined and evaluated, the next step is to select a method to recover the uranium. The method chosen must be physically, economically, and environmentally compatible with the characteristics and location of the uranium ore body. Factors affecting the choice of recovery method include: spatial characteristics of the ore body (size, shape, attitude, and depth); physical properties of the ore body and the surrounding rock (structural support of underground operations); groundwater and hydraulic conditions; economics (ore grade, expected mining costs, and planned production rates); and environment (surface disturbance and reclamation, preservation of air and water quality) (Battelle, 1974).

The uranium industry uses a variety of methods to recover uranium from the ground, but these basically fall into two categories: conventional and nonconventional. Conventional methods include underground and open-pit mining; nonconventional methods include solution mining (in-situ and heap leach), recovery from uranium mine water, and uranium recovery as a by-product of phosphate and copper operations. As shown in Table 21.1, conventional recovery methods presently account for 92 percent of U_3O_8, with underground mines yielding 80 percent and open-pit mines 12 percent. However, the annual production from one average open-pit mine is much greater than from one average underground mine.

Open-Pit Mining

Although similar to open-pit (strip) coal operations, open-pit uranium mines usually cover much less area and may go deeper (Hebb and Morse, 1976). The open-pit method is favored when deposits are near the surface and are covered with loose, easily-removable soil. Pit depths range from just a few feet to as

Plate 21.1 Surface support facilities for the underground Schwartzwalder Mine in Jefferson County, Colorado. (Courtesy of the Cotter Corporation)

TABLE 21.1

DISTRIBUTION OF 1977 U_3O_8 PRODUCTION SOURCES BY MINING METHOD

Source	Number*	% of Total
Underground Mines	251	80
Open Pit Mines	36	12
Others: Heap Leach, Mine Water, Solution Mining, Low-Grade Stockpiles	27	8
Total	314	100

SOURCE: U.S. Department of Energy, 1978a, p. 17.

*Includes shipments from 287 operating mines and 27 miscellaneous sources.

much as 300 feet, with the overburden ratio (thickness of material covering the ore to the ore body thickness) varying from 8:1 to as much as 35:1. The latter, a higher ratio than usual for other minerals, is justified by the greater product value of uranium (Clark, 1974).

The size of the mining operation determines the type of earth-moving equipment used. Normally scraper loaders, bulldozers, and gas and diesel shovels or rippers are used, but smaller ore bodies can even use backhoes economically. Many open-pit mines suffer groundwater intrusion and require pumping of water from the pit to maintain a workable floor surface. This water, subject to EPA standards, is allowed to seep into the ground away from the mine or to drain into nearby creeks or rivers.

Underground Mining

Again the techniques are similar to those described for coal mining (Hebb and Morse, 1976), but there are important differences related to seam sizes and mine ventilation systems. For example, most uranium seams are long, thin, and erratic in occurrence requiring working equipment that can be readily moved from site to site. The largest uranium ore bodies mined underground are about one-half mile long, several hundred feet wide, from five to a hundred feet thick, and lie several hundred feet below the surface.

One of uranium's decay products is radon-222, a radioactive gas. To prevent this gas from concentrating and exposing the

Plate 21.2 Underground hardrock uranium mining at the Schwartzwalder Mine, Jefferson County, Colorado. (Courtesy of the Cotter Corporation)

miners to dangerous levels of radioactivity, uranium mines must have special ventilation systems. Large-capacity pumps force fresh air down the main shaft, and there are special exhaust shafts at tunnel ends. These are designed to maintain radon at EPA-approved levels. (See General Radiation Hazard, Chapter 24, for additional details.)

NONCONVENTIONAL RECOVERY METHODS

The uranium industry is testing alternative methods of uranium recovery to counter the rising capital and operating costs associated with underground and open-pit mining. Techniques being investigated include solution mining (heap leach and in-situ); bacterial leaching; hydraulic mining; and by-product recovery from phosphate and copper operations. Uranium production from such nonconventional methods was approximately 500 tons of U_3O_8 in 1976, and could reach as much as 3,000 tons in 1978 (Klemenic, 1976, p. 251).

In-Situ Solution Mining

In-situ solution mining evolved from heap leach (percolation) mining. Reduced to its essentials, and shown conceptually in Figure 21.1, in-situ mining (also called in-place leaching) uses an acid or an alkaline solution, pumped into an underground ore body, to dissolve the desired components. The fluid is then pumped to the surface where uranium is removed and the solution, if alkaline, is recycled. To use this method, however, the following geologic conditions must exist:

1. The ore must be below the groundwater table in a sandstone aquifer between two impermeable beds (as shown in Figure 21.2) so the leaching solutions can be confined to the ore vicinity;
2. The horizontal permeability of the ore-bearing sandstone should be enough to allow sufficient leachant to contact the ore;
3. The ore beds must have either a high ratio of horizontal to vertical permeability or occupy at least 10 percent of the aquifer to prevent uneconomic dilution of the leachants; and
4. At a price of $40 per pound the deposit must contain at least 200,000 pounds of ore with an average grade of better than 0.075 percent U_3O_8. (Others have stated 0.05 percent--author's note.) (Westphal, 1978, pp. 92-93.)

The leaching process is performed entirely underground by drilling inflow wells for injecting solvent and production wells for recovering liquor as shown in Figure 21.2. The spacing of wells depends on the dimensions of the ore body, as determined by exploratory drilling. Injection of the solvent,

Solution Mining (Conceptual)

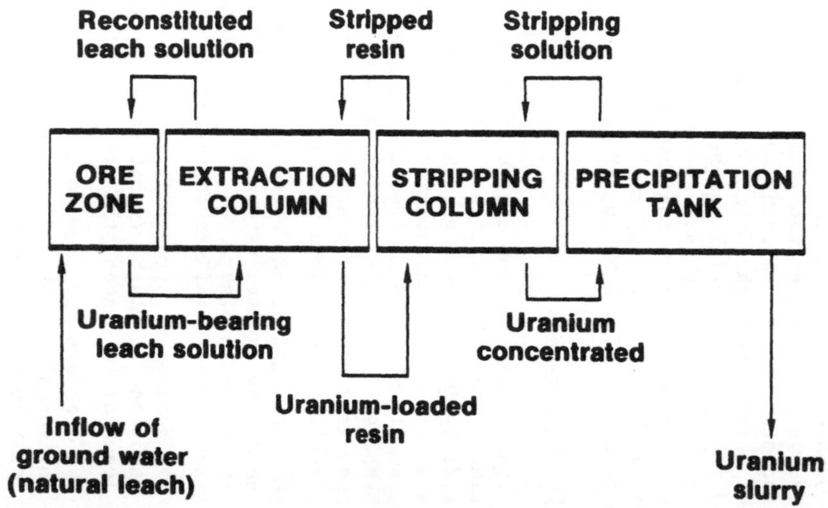

Figure 21.1

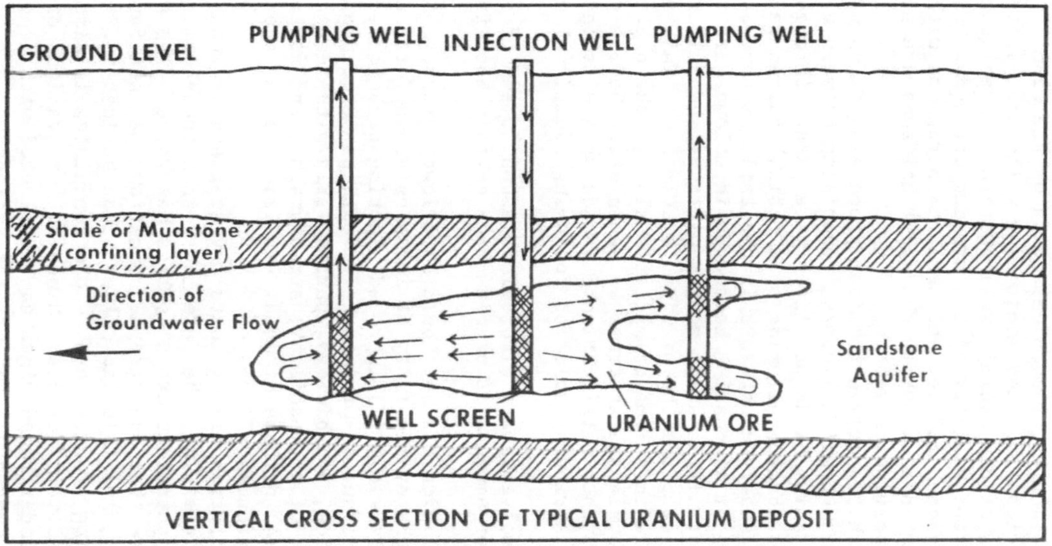

Figure 21.2 In-situ Solution Mining

Source: Westinghouse Electric Corp., 1972.

a dilute sulfuric acid containing additive oxidants, or a dilute carbonate solution if the host formation permits, continues until the concentration of uranium in the leach solution is reduced to a level indicating depletion. Using a pattern with three to five wells in simultaneous operation, zone depletion may take several months or more depending on the size of the ore body. Additionally, many monitoring wells are located around the edges of the overall well pattern. These keep leachant confined to the vicinity of the uranium ore body. Uranium recovery can approach 100 percent, except in regions with multiple uranium horizons where the upper horizons are not underlain by impervious layers.

One approach to surface chemical processing is using acid solvents. Uranium-bearing liquors, containing between 0.1 - 0.30 grams of uranium oxide per liter, are passed through ion exchange beds to absorb the oxide on a resin. Barren leach solutions are discharged as waste into a tailings pond allowing solvent recovery or disposal. Loaded resin columns are treated with a mixture of nitric acid, sodium nitrate, and sulfuric acid, to desorb or place the uranium in solution. This eluate is precipitated with a lime or magnesia slurry; the precipitate is allowed to settle and separate by decantation before shipment to a mill for further purification. The decantate is then recycled to the stripping solution following additional nitric acid makeup.

Carbonate solvents are also being tested to see if they will yield higher amounts of uranium from certain host formations.

In-situ solution mining should have lower production costs because of (1) the availability of inexpensive sulfuric acid (primary solvent used); (2) improved blasting techniques; (3) better understanding of interrelationships between mineralogy, chemistry, physics, and biology, particularly with respect to the complexities of permeability, stratigraphy, and formation chemistry; and (4) perhaps of equal significance, a potential for minimal environmental impact.

Although solution mining costs may approach or even exceed those of open-pit mining, they are lower than those of underground mining; operations are also less hazardous. The primary disadvantages are that underground conditions may not favor maximum recovery, and injected solvents can be lost in fractures, thus contaminating groundwaters. Solution mining is also suited to "secondary recovery" of ore. Mined-out excavations or stopes and tunnels in underground mines can be flooded with leach solutions, then pumped out for uranium extraction.

Several companies have been testing solution mining in Wyoming and Texas, with plans for commercial uranium recovery. The largest plant is rated at 500,000 pounds per year. They expected to recover 700,000 pounds of U_3O_8 in 1977 by solution mining (Davis, 1977). The figures on the actual amount recovered are not yet available.

Plate 21.3 Solution Mining. Intercontinental Energy Corporation's Zamzow Uranium Mine in Live Oak County, Texas, designed to produce 300,000 pounds of U_3O_8 annually. Solution mining in Colorado would use similar surface equipment and facilities. (Courtesy of Behrent Engineering Co.)

Heap Leaching

Often stockpiles of low-grade ore may be processed profitably by heap leaching. The ground at the site is graded to a smooth, sloping surface and covered with a thin polyethylene sheet. Piles of crushed ore are placed on the sheet and sulfuric acid solutions are allowed to percolate slowly through the ore. The solutions are collected and processed, as described under solution mining, leaving the ore with a residue of generally less than 0.05 percent uranium. Western Nuclear's Day Loma facility produced some 12,000 pounds of uranium per month using this method (Clark, 1974).

Heap leaching operations in Colorado are being performed at Naturita. The facility here was originally engineered in 1977 to treat 600,000 tons of existing tailings. Beginning production in December 1977, it is currently producing 1,000 pounds per day of yellowcake and 4,600 pounds per day of vanadium (Mining Record, 1978, p. 1). The contractor, Ranchers Development and Exploration Corporation, has applied to the State for the necessary licenses and permits to modify the design of the Naturita facility and move it several miles to a site near Durango where an additional 1.4 million tons of existing tailings will be processed at about the same daily rate (Montgomery, 1978). The contractor expects the Durango facility to be on line in the fall of 1979.

Union Carbide is licensed to heap leach a low-grade uranium ore and tailings. A listing of Colorado processing operations is given in Table 22.2 (See Processing, Chapter 22).

Bacterial Leaching

The value of this method is that it converts uranium into its highest oxidation state which is soluble in acid, making uranium more readily extractable. Pyrite (FeS_2) is often found with uranium ore and, in the presence of water, pyrite converts to ferrous sulfate and sulfuric acid. Certain acid-tolerant bacteria are capable of oxidizing ferrous sulfate to ferric sulfate, which, in turn, oxidizes the uranium to its most readily soluble form. The amount of sulfuric acid needed is reduced from eighty pounds per ton (for solution mining) to twenty-five pounds per ton (Clark, 1974).

Although successful, the bacterial leaching process takes a long time. It is used chiefly on worked-out mines and on low-grade materials above and below ground. Its chief drawbacks are a lack of enough pyrite in the uranium ore body or too much neutralizing calcium carbonate, which will not support bacterial activity.

Hydraulic Mining

Remote mining, using a high pressure water jet at the bottom of a cased drill-hole, creates a slurry that is pumped

Plate 21.4 Facilities to recover uranium from low-grade ore at Rifle, Colorado. (Courtesy of the Union Carbide Corporation)

to the surface for conventional leaching. Hydraulic mining is useful for small, isolated ore bodies. A pilot test in a sandstone environment, a joint venture of the Marcona Corporation and the Rocky Mountain Energy Company, is currently in progress. The disadvantages of this method are the relative lack of selectivity, incomplete recovery, and blockages caused by hard ribs in the ore body. Cost savings and reduced environmental impact, however, may be significant advantages for this method (Davis, 1977).

Uranium Mine De-watering

In selected cases, uranium may be recovered from the water removed from a uranium mine. Recovery of uranium mine water amounted to approximately 100 short tons of U_3O_8 in 1976 (Facer, 1976, p. 200). This amount is expected to increase as more underground (wet) mines come into production in the future, but is unlikely to account for more than 1 or 2 percent of the total United States uranium production.

Recovery from Phosphate and Copper Operations

Phosphate rock normally contains a low concentration of uranium, about 0.01 percent or less. In the conversion of the phosphate mineral into agricultural fertilizer, dissolved uranium exists as a residual product, and can be recovered by chemical processing. The many tons of phosphate rock treated can therefore generate a small, but significant, quantity of uranium oxide (U_3O_8). While confined largely to Florida and North Carolina, uranium may eventually be recovered from deeper phosphate bodies in Idaho and the Pacific Northwest.

Leach solutions from western copper mine dumps frequently contain two to twenty parts of U_3O_8 per million parts of solution (Facer, 1976). While this concentration is less than the pregnant leach solutions from in-solution mining operations, which average twenty-nine parts per million (Frank, 1976), the recovery process is technically feasible. Its economic feasibility has yet to be proven. One plant, designed to recover uranium from copper dumps at Kennecott's Bingham Canyon (Utah) mine, was completed by the Wyoming Mineral Corporation in October 1977. Production from this and other such facilities could reach 500 - 1,000 tons of U_3O_8 by 1983 (Facer, 1977a).

Future Mining Capacity

After 1985, most uranium will probably be produced by major underground mines, principally in New Mexico and Wyoming. Average depths considered are 2,000 feet or more, with one of 3,500 - 4,200 feet in New Mexico in the planning stage. These mines will need long lead times for shaft-sinking and high-cost ventilation systems. Wyoming mines will be less deep (600 - 2,000 feet), but will suffer the difficulties of working generally in unconsolidated tertiary sediments, and will require further developments in mining technology (Koch, 1977).

Solution mining, or in-situ extraction, will play a larger role in the uranium industry. This optimistic view is not shared by everyone, but as experience is gained, its role could grow markedly (Schock, 1977).

22
Uranium Processing

INITIAL PROCESSING (MILLING)

In conventional mining operations processing of uranium ore begins at a uranium mill. The milling process converts ore with a U_3O_8 content of 0.1 - 0.2 percent into a compound called "yellowcake." Yellowcake contains approximately 80 - 95 percent U_3O_8. The specific process, or circuit, used varies from mill to mill depending upon the presence of undesirable materials in the ore and any valuable by-products that can be recovered along with the uranium. Table 22.1 shows a generalized process flow (mill circuit), and is illustrated in Figure 22.1.

The simultaneous recovery of by-products and uranium shifts the economics of processing and can result in favorable costs that enable recovery of lower grade uranium ores. Economic feasibility of by-product recovery depends upon what and how much co-exist with the uranium. In Colorado, the most common metal associated with uranium is vanadium. Other elements found with uranium, in Colorado and elsewhere, include molybdenum, rhenium, selenium, copper, silver, scandium, arsenic, mercury, thorium, and radium (McGinley and Facer, 1976). Table 22.2 lists the present and planned uranium mills in Colorado. The two Union Carbide mills, at Uraven and Rifle, recover vanadium; the Cotter Corporation's mill at Canon City recovers copper, cobalt, and nickel.

While uranium ores generally show wide variations in composition, most mill processes are designed to accommodate a limited range of feed materials. Processing requirements reflect the grade of ore (uranium content), co-existence of other recoverable minerals with uranium, and the economic and environmental restrictions of the milling process itself. The percentage of lower grade uranium ores shipped to uranium mills will increase in the future, as shown in Figure 22.2. Consequently, the uranium industry must improve unit operations, in a chemical/metallurgical engineering sense, and should assess energy requirements for each potential processing system.

Steps Common to Most Uranium Processing (milling) Methods

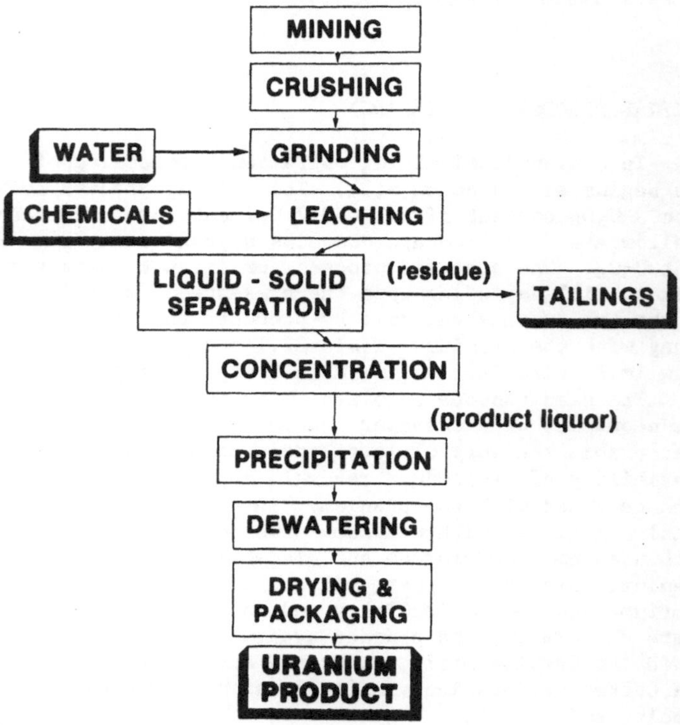

Figure 22.1

Plate 22.1 Uranium milling facility at Uravan, Colorado. (Courtesy of the Union Carbide Corporation)

MILLFEED GRADES AND MILL RECOVERIES

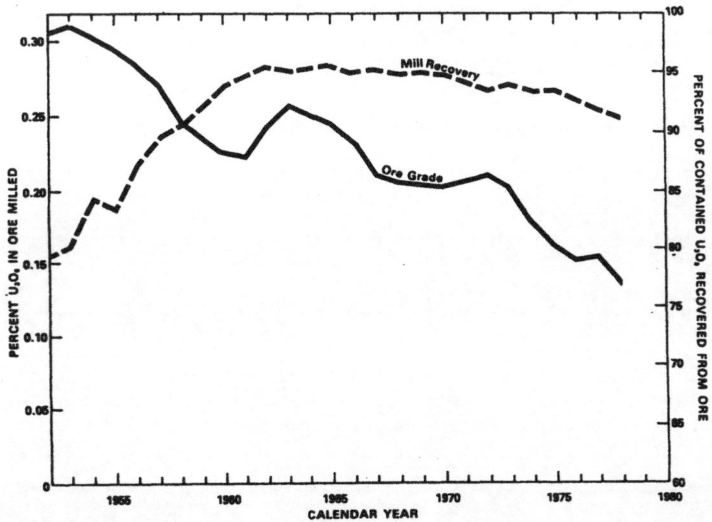

Figure 22.2

Source: Facer, 1978, p. 8.

TABLE 22.1

URANIUM MILLING PROCESS FLOW

Sorting	Blending is done before delivery to the mill to provide ore having uniform physical characteristics and uranium content. This step assists in achieving an optimum process flow.
Preconcentration	Beneficiation and upgrading appear desirable, but economic tradeoffs are necessary. The latter are ore- and process-dependent.
Crushing and Grinding	Reducing particle sizes enables rapid chemical reactions. Jaw crushers (15-40 inches) reduce sizes to 1/4 to 1 inch. The ore is carried by belt-type feeders to rod or ball mills for grinding to minus 28-mesh (acid leach) or minus 65-mesh (alkaline leach). Water is added in the grinding step for acid-carbonate-bicarbonate for alkaline.
Leaching	Removal of uranium minerals from the host rock by bringing them into solution is called leaching. The primary leaching agents are acid (sulfuric acid; 80% of United States yellowcake is produced using this material) and alkaline (sodium carbonate and bicarbonate). The choice is a function of the physical characteristics of the ore: type of uranium mineralization, ease of liberation, and other constituents present. Quantity of limestone present is dominant.
Liquid-Solid Separation	The leached solution is separated from the barren solids. Countercurrent decantation or filtration uses water to aid the separation.
Concentration and Purification	This step involves separation of uranium from other leached minerals, using either ion-exchange resin or organic solutions, followed by extraction of the uranium with water soluble salts. Countercurrent techniques are used in the organic solvent systems, using a series of extraction tanks. This is the most complex step in the process as large quantities of solvents, emulsion

TABLE 22.1 (continued)

	formation, etc., are cost-related problems. The alkaline process requires no purification at this point.
Precipitation	The addition of ammonia, sodium hydroxide or sodium peroxide at elevated temperatures leads to the formation of insoluble U_3O_8 or yellowcake.
Drying and Packaging	The wet precipitate is dewatered by thickening, followed by filtration, with the filtercake dried at 700-800°F. The dried product is then crushed and packaged in drums for shipment to refineries where pure uranium or one of its compounds is prepared.
Secondary Metal Recovery	In many instances, other metals are present in uranium ores in sufficiently high concentrations to warrant economic extraction.

SOURCE: author

Although milling costs per ton of ore processed are expected to remain roughly stable in the immediate future, they will probably increase over the long-term, since the milling cost per pound of uranium ore processed varies directly with uranium content (Hazen, 1978, p. 120). On the other hand, larger through-put may reflect economies of scale and help stabilize milling costs.

Process variations at United States open pit and underground uranium mines are contrasted with solution mining in Table 22.3. Solution mining presents a significant reduction in steps when compared with traditional mining methods. Such reductions normally reflect lower cost operations (See Commercial Development, Chapter 23), but not all underground ore deposits are amenable to solution mining. Nonetheless, development experiences with this technique may well point to expanded uses.

Returning to milling operation, DOE projected a requirement for some 40,000 tons of yellowcake in 1985 and an estimated production of about 30,000 tons (ERDA, 1977b, pp. 14, 17). If 30,000 tons were produced from ores averaging 0.15 percent U_3O_8, about 200 million tons of ore will require milling. At about 300 days per year, the daily milling rate will be about 70,000 tons. This is about twice the existing milling capacity of the United States at this writing, indicating a serious milling capacity shortfall by 1985, although some small percentage will be absorbed by solution mining.

TABLE 22.2

COLORADO URANIUM PROCESSING OPERATIONS

Mills	Company	Site	Capacity (tons ore/day)
Operational	Union Carbide	Uravan	1,300
	Cotter Corp.	Canon City	450
	Union Carbide	Rifle	---(a)
Planned	Cotter Corp.	Canon City	1,000(b)
	Cyprus/Wyoming Mineral	Tallahasse Creek Area	Several 1,000 (c)
	Pioneer/Uravan Corp.	Slickrock	1,000(d)
	Boulder Mining	Central City	small
Heap Leaching			
Operational	Union Carbide	Maybell	N.A.
	Ranchers	Naturita (600,000 tons tailings)	1,000
Planned	Ranchers	Durango (1.4 million tons tailings)	1,000

SOURCE: Montgomery, 1978.

(a) On a vanadium circuit only; may reactivate for uranium in 1980.
(b) Construction virtually complete; licensing decision by December, 1978.
(c) Projected for early 1980s.
(d) Early estimate; no projected operational data available.

TABLE 22.3

PROCESS VARIATIONS AT U.S. URANIUM MINES AND MILLS

Open Pit Mines	Underground Mines	Solution Mining
Development drilling	Development drilling	Development drilling
Stripping*	Shaft sinking*	
Mine Waste*	Development drifting	
Waste dump*	Waste dump*	
Develop ore faces	Develop stopes	Drill wells
Drill, blast*	Drill, blast	Leach uranium
Load	Muck out	
Haul	Haul	
	Hoist	
	Haul to mill	Pump solution to mill
Crush*	Crush*	
Grind	Grind*	
Leach uranium	Leach uranium	
Liquid-solid separation	Liquid-solid separation	Liquid-solid separation
U_3O_8 concentration	U_3O_8 concentration	U_3O_8 concentration
Precipitate, dry, package	Precipitate, dry, package	Precipitate, dry, package
Tailings dam* operations	Tailings dam* operations	Recirculate leach solution
Reclamation	Reclamation	Aquifer restoration
16**	18**	9**

SOURCE: Environmental Protection Agency, 1978, Section 3-4.

*Denotes stages generally producing significant changes in or affecting land surfaces, water quality, personnel safety or radiation exposure.

**Total stages to produce a saleable product.

SUBSEQUENT PROCESSING - CONVERSION AND ENRICHMENT

In order to convert yellowcake (U_3O_8) to a form that can be used as nuclear reactor fuel, further processing is necessary. Basically, this processing takes place in two stages: first, conversion to highly purified uranium hexafluoride (UF_6), a gaseous compound; and second, concentration of the U-235 component of the gas. There are several conversion plants in the United States, but none are in Colorado.

The two UF^6 production processes currently used are the dry hydrofluor process and the wet-solvent extraction-florination process. At present two plants are in operation (one of each type), each producing about equal quantities of UF^6 (University of Oklahoma, 1975).

The next step is enrichment (concentration) of the U-235 component of the UF_6, from its naturally occurring concentration of only 0.7 percent to a much greater percentage. Light Water Reactors (LWR) require a fuel that is approximately 3 percent U-235, while the high-temperature, gas reactor (HTGR) requires a U-235 concentration of approximately 95 percent. Currently, there are three operational enrichment plants in the United States: Portsmouth, Ohio; Paducah, Kentucky; and Oak Ridge, Tennessee.

FUEL FABRICATION

Pure gaseous UF_6, enriched to about 3 percent U-235, is chemically reduced to uranium dioxide (UO_2) or metallic uranium, for use as reactor fuel. The chemical form of the material used depends on the type of reactor. LWRs use the dioxide compressed into cylindrical ceramic pellets. The usual configuration of a single fuel element consists of a thin-walled, stainless steel (or zirconium) tube packed with UO_2 pellets in close contact with the fuel element wall, giving maximum heat transfer from the fissioning pellet to the wall itself. Tubes are sealed by welding in an inert gas atmosphere, such as helium, under slight pressure, to permit leak testing before final assembly and to aid in heat transfer from the fissioning uranium to the fuel element wall.

Fuel elements, normally about eight feet long, are then packaged in a metallic grid-type assembly which holds each element in a rigidly fixed position during its fuel life in a reactor. Position, or spacing, allows water to flow between the tubes, performing neutron-moderating and heat transfer roles in LWRs. Reactor type and output power determine the amount of uranium required and hence, the number of fuel elements and their configuration, defined as a reactor fuel core or fuel assembly. (See Appendix D for further information on nuclear reactor fundamentals.)

23 Commercial Development Considerations

The commercial development of uranium production facilities (mines and mills) in the United States is influenced principally by demand for uranium, and not the available supply, as given by the estimated ore reserves. That is, given an existing supply of uranium, there are technical, political, and economic factors which interact to encourage or discourage its extraction and use.

Uranium is unique in that it is used primarily for one purpose: nuclear fuel. If National Defense requirements are ignored, the principal demand for uranium stems from its use in nuclear power plants to generate electricity. The amount of uranium required to fuel and operate a nuclear reactor is known with a relatively high degree of certainty. However, estimates of the number of nuclear reactors that will be built and operated in the future are rather uncertain. The decision by a utility to build a nuclear power plant, as opposed to a fossil fuel, geothermal, or solar facility, is based not only upon economic considerations, but also upon other factors such as public acceptance, reliability, safety, leadtime, etc.

Historically, electricity generated by a nuclear plant has been less expensive than that from a fossil-fuel plant. This occurred simply because, over the lifetime of the plant, the low nuclear fuel cost incurred in operating the nuclear power plant more than offset the high initial capital costs (Iowa Energy Policy Council, 1975, p. 31). However, there are other factors which tend to discourage the building of nuclear power plants. These include: environmental constraints; long regulatory delays; capital cost uncertainties (especially inflation); capital shortages (including resistance to rate increases); construction delays (due to lack of skilled labor or the late delivery of precision parts); and public acceptance. Even after a nuclear power plant is built, the uranium requirement for the lifetime of the plant will vary according to: the type of nuclear reactor, the tails assay of the uranium enrichment plant, and whether or not uranium and/or plutonium will be recycled from the spent fuel. (For a more detailed discussion of these factors, see Technology Developments Affecting Uranium Needs, Chapter 27).

Interestingly enough, while an increase in the price of uranium provides a strong incentive for uranium exploration and development, it has less effect on the cost of the fuel rods used in a nuclear reactor. The cost of processing yellowcake (80-95 percent U_3O_8) into fuel rods was estimated as approximately \$100 in 1975 (conversion to UF_6, enrichment, fabrication, and recovery). At this cost, if the price of yellowcake were to increase 400 percent, e.g., from \$25 per pound to \$100 per pound, the result would only be a 60 percent increase in the cost of a fuel rod (from \$125 per pound to \$200 per pound), and the price of the energy content would still be \$1 per million Btus (SINB, 1976, p. 70). This contrasts very favorably with similar costs for coal, natural gas, and oil at today's prices.

Aside from uranium price considerations (next section), another major influence affecting the commercial development of a uranium property is the long lead-time associated with getting it into production. As shown in Figure 23.1, the expected lead time from the beginning of exploration to the initiation of production may be ten years. This estimate implicitly assumes that environmental baseline data will be collected, and an environmental impact report will be written and approved, before development of the mine site and construction of the mill. This means that a relatively large cash flow and a rather high rate of return will be required to recover the monies spent early in the project on exploration, land acquisition, and development.

URANIUM PRICE CONSIDERATIONS

The uranium market has never been what economists term "pure competition," that is, a marketplace with many buyers and sellers, each of whom operates independently. Instead, in the 1950s and early 1960s, the Federal government (Atomic Energy Commission) was the only uranium purchaser. In fact, during this period, the AEC encouraged development of the United States uranium industry by guaranteeing to buy uranium at a fixed price, thus insuring that National Defense requirements would be met. This incentive worked so well that there was a glut of uranium on the market by the early 1960s. As a result, production dropped by nearly 50 percent. Current production levels (Table 23.1) are still lower than they were in 1959-1962. More recently, due to the increased demand for uranium by utilities, the uranium market has become a "seller's market." However, this situation may change in the near future, as shown in Table 23.2.

Apparent availability of uranium to United States buyers is shown in Column (7) by summarizing domestic purchase commitments of (1) domestic and (2) foreign uranium, and (3) additional domestic U_3O_8 to be available for sale. This supply would be

Steps in Uranium Resource Development

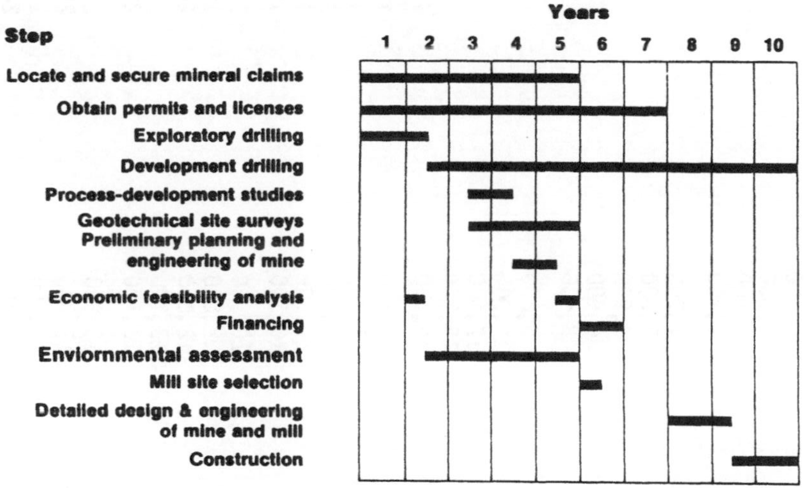

Figure 23.1
Source: Environmental Protection Agency, 1978, p. 3-4.

TABLE 23.1

URANIUM ORE PRODUCTION, 1947-1976 (Tons U_3O_8)

Year End	Shipment to Mills*	Cumulative Production
1947	-	-
1948	100	100
1949	500	600
1950	800	1,400
1951	1,100	2,500
1952	1,300	3,800
1953	2,300	6,100
1954	3,500	9,600
1955	4,400	14,000
1956	8,400	22,400
1957	9,800	32,200
1958	14,000	46,200
1959	17,400	63,600
1960	18,800	82,400
1961	18,500	100,900
1962	17,100	118,000
1963	14,700	132,700
1964	13,900	146,600
1965	10,600	157,200
1966	10,100	167,300
1967	10,900	178,200
1968	12,800	191,000
1969	12,600	203,600
1970	13,100	216,700
1971	13,100	229,800
1972	13,900	243,700
1973	13,800	257,500
1974	12,600	270,100
1975	12,300	282,400
1976	14,000	296,400
1977	16,700	313,100

SOURCE: U.S. Department of Energy, 1978a, p. 9.

*Includes miscellaneous U_3O_8 receipts from mine waters, heap leach, solution mining, and refining residues.

TABLE 23.2

U.S. URANIUM SUPPLY AND MARKET SUMMARY

Years	Domestic Production (1) Sales Commitments To Domestic Buyers	(2) Sales Commitments To Foreign Buyers	(3) Estimated U_3O_8 To Be Available for Sale	(4) Procurement of Foreign Uranium	(5) Reported Unfilled Requirements	(6) Total Domestic Production	(7) Total* Domestic Supply	(8) Apparent Buyer Requirements
						1+2+3	1+3+4	1+4+5
1978	19,100	1,500	3,100	1,600	300	23,700	23,800	21,000
1979	17,700	1,400	4,100	1,600	1,600	23,200	23,400	20,900
1980	19,600	1,000	5,000	2,700	3,000	25,600	27,300	25,300
1981	19,600	400	8,200	3,600	5,700	28,200	31,400	28,900
1982	19,500	200	10,500	3,600	8,600	30,200	33,600	31,700
1983	17,000	200	14,000	3,300	8,000	31,200	34,300	28,300
1984	13,000	200	16,300	3,100	12,400	29,500	32,400	28,500
1985	11,700	200	16,900	2,900	14,100	28,800	31,500	28,700
1986	9,100	200		1,750	19,500			30,350
1987	8,800	200		1,750	23,300			33,850
1988	7,900			1,750	24,700			34,350
1989	7,500			1,750	28,100			37,350
1990	6,400			1,750	28,600			36,750

SOURCE: U.S. Department of Energy, 1978b, p. 23.

*Buyers have an additional 28,700 tons U_3O_8 in inventory.

augmented by the 28,700 tons of buyer inventories. Apparent buyer uranium requirements, Column (8), is obtained by summing purchase commitments--(1) domestic and (4) foreign--and (5) unfilled requirements.

Comparisons of the various data in Table 23.2 indicate that there is sufficient uranium supply to take care of buyers' perceived demands, at least through 1985. However, as pointed out previously, unfilled requirements are probably understated, and the supply computation in Column (7) assumes that all additional domestic U_3O_8 for sale would be made available to domestic buyers and not exported. Further supply expansion will depend on additional industry resource and production efforts which will, in turn, be influenced by future uranium demand developments (U.S. Department of Energy, 1978b, p. 22).

The Westinghouse Corporation has alleged that the current seller's market condition is artificial, created by a foreign-based uranium cartel; the case is currently in litigation. However, the price increase is more probably due to sharp increases in operating equipment and supplies, especially fuels, a sharp increase in the demand for uranium by new and existing power plants, and a continuing decrease in ore grade. Regardless of the cause, the price per pound of uranium (U_3O_8) more than doubled between 1973 and 1976--up from $7.10 per pound to $16.10 per pound (ERDA, 1977a, p. 97). In spite of this, there are several reasons why producers are not rushing to expand existing production operations nor are they making large immediate commitments to new uranium exploration and development. Aside from lead time considerations, both existing and potential uranium producers are rather cautious, having either experienced or observed the problems of an over-supplied market. While there are more than thirty companies in the business of mining and milling uranium, common holdings and joint ventures reduce this number to about twelve. Only three of these are solely dependent upon uranium mining or milling for cash flow; the others are primarily engaged in the production of oil or gas or extraction of other minerals (Battelle, 1974, p. 4.12). To encourage uranium mining and/or milling ventures, profit margins must be sufficiently high and economic risk (uncertainty) sufficiently low, when compared with other energy or mineral projects competing for the same investment dollar. Unfortunately, this is not often the case.

Part of the uncertainty of the uranium market is due to the lack of long-term contracts (Figure 23.2) (U.S. Department of Energy, 1978b, p. 4). Approximately 135,000 short tons of U_3O_8 have been contracted for delivery from 1978 to 1985; projected requirements are over one and one-half times that amount. (See Table 23.2.)

Another element of uncertainty stems from confusion regarding the terms "average contract price" and "market price." "Average contract price" is the contract price for the bulk of uranium deliveries and is weighted by the quantity sold at that price. (It might better be termed the "median price.")

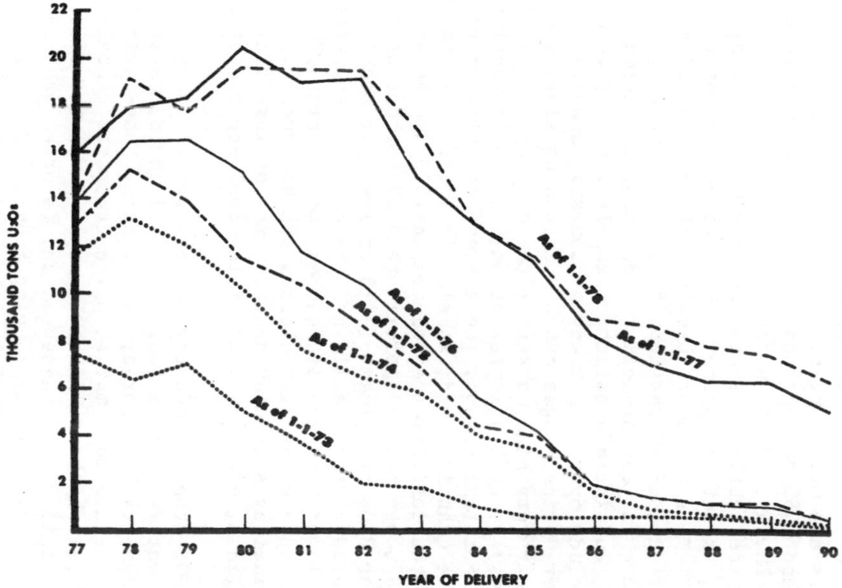

Figure 23.2 Domestic Uranium Commitments to Domestic Buyers
Source U.S. Department of Energy, 1978 b, p. 4.

"Market price" is not a contractual but a "spot sales" price. Consequently, "market price" tends to be high, due to a limited supply of uranium for such sales, but quantities involved are usually small.

In the last several years, use of "market price" contracts has increased considerably. In this type of contract, prices are determined at or sometime before time of delivery based on prevailing prices. Figure 23.3 shows the range of uranium contract prices and market price settlements for deliveries between 1977 and 1985 as reported in a 1/1/78 survey (U.S. Department of Energy, 1978a, p. 74).

Market price settlements are included with "contract prices" since, as settled prices, they are equivalent to "contract prices." This procedure provides the best available overall average price (U.S. Department of Energy, 1978b, p. 7). Prices shown are in year-of-delivery dollars, as estimated by the purchaser, and include estimated escalation. Price data for most years include over 80 percent of domestic uranium requirements. However, price ranges can be misleading since the extremes (either very high market prices or very low, earlier contract prices) generally represent only a small fraction of total sales. For this reason, the average contract price, weighted by the quantity sold, is also indicated. Approximately 15 percent of the deliveries scheduled in 1978 are at prices in excess of \$20 per pound; approximately 50 percent of deliveries in 1980 are for U_3O_8 at prices of \$15 or less per pound. Thus, many uranium producers are still held by contract to prices significantly lower than the oft-quoted "market price." Additionally, even when escalation clauses can be invoked, any price increase obtained by a uranium producer may be insufficient to offset declining ore grades or rising fuel and processing costs.

In past DOE uranium industry survey reports, only "market prices" and "contract prices" were included and discussed. However, in the January 1978 survey, a new term, "base price" has been included. Since most market price contracts contain base (or floor) prices which provide a lower limit on the eventual settled price, this survey requested and reported data on base prices. Average base prices, ranging from \$34.30 in 1978 to \$54.55 in 1985, are approximately 2 to 2.5 times as high as the average prices reported in Figure 23.3, and are closer to prices for new procurement (U.S. Department of Energy, 1978b, p. 7).

More price uncertainty results from expected imports and exports of U_3O_8. As one means of supplementing domestic uranium supplies, ERDA has allowed (beginning in 1977) utilities to satisfy up to 10 percent of their uranium requirements with imported uranium. (Before 1977, ERDA enrichment facilities would only accept domestically-produced uranium.) This allowable percentage will increase each year until it reaches 100 percent in 1984 (Chemical and Engineering News, 1977, p. 6). Estimates are that approximately 10 to 20 percent of United States uranium

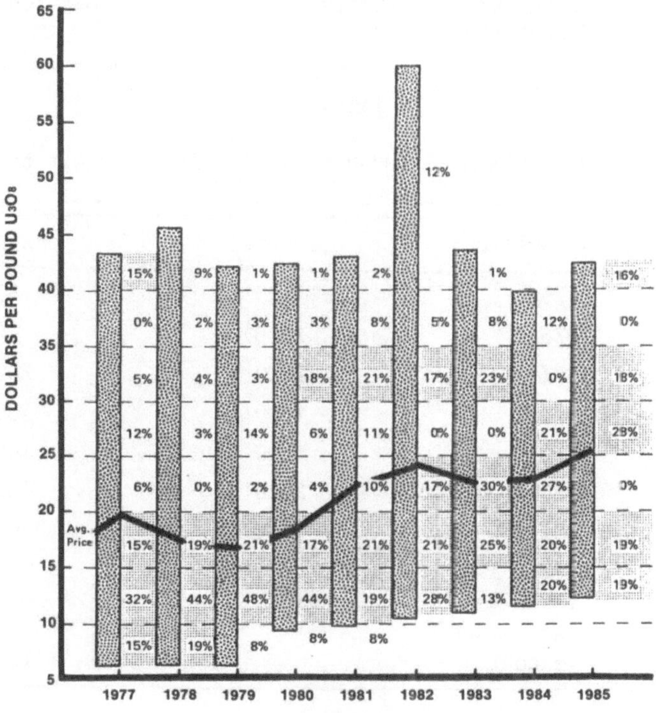

Figure 23.3 Distribution of U_3O_8 Prices, January 1, 1978 Contract Prices and Market Price Settlements

Source: U.S. Department of Energy, 1978 b, p. 10.

requirements will be imported during the next decade, but that substantial increases over this amount are unlikely due to rapidly rising demand for uranium elsewhere throughout the free world (Engineering and Mining Journal, 1977, p. 92). Export of domestic uranium increases the marketing area available to United States uranium producers but reduces the supply available to domestic users. Unfortunately, aside from scheduled export sales reported by ERDA, no other export data are available that indicate future export activity. Table 23.3 summarizes uranium imports and exports.

TABLE 23.3

UNITED STATES IMPORTS AND EXPORTS OF U_3O_8 (tons)

Year of Delivery	Imports	Exports
1975	700	500
1976	1800	600
1977	2800	2000
1978	1600	1500
1979	1600	1400
1980	2700	1000
1981	3600	400
1982 - 1990 (imports)	21,600	
1982 - 1988 (exports)		1200

SOURCE: U.S. Department of Energy, 1978a, pp. 75-76.

The last uranium price consideration concerns the relationship between ERDA's "forward cost" category and the ore grade necessary to make a uranium mine a long-term paying proposition. In determining uranium ore reserves for a particular property, at a selected maximum allowable cost per pound, ERDA first determines a cutoff grade equal to the operating cost, and then adds an expected cost per pound of U_3O_8 to cover forward capital costs; that is, those costs that would be incurred to develop the property, including construction of required facilities. (Cut-off grade defines the lowest grade material that will be mined; material grading less than this will be left unmined in the deposit.) If the resultant cost per pound of U_3O_8 is less than the specified maximum allowable cost per pound, then the material is considered as ore reserve at the stated forward cost. The actual average grade of material extracted from a particular ore body will exceed the cutoff grade by some amount; the actual amount depends upon the geological characteristics of the ore deposit and the mining method utilized. This "excess" ore value is used to recover "sunk" costs (costs previously incurred for exploration, land acquisition, interest, etc.), to pay income taxes, and to assure a reasonable profit.

Figure 23.4 illustrates the relationship between these various cost recovery categories and the grade of ore, for an assumed maximum allowable cost of $15 per pound of U_3O_8. In this example, operating costs are recovered if the ore grades 0.09 percent or higher. The minimum average grade that will recoup all forward capital costs, as well as the operating expenses, is presently computed at 0.13 percent U_3O_8. As a matter of fact, for the example given, the average ore grade for the entire deposit is 0.17 percent, when a cutoff grade of 0.09 percent is specified. This average grade of 0.17 percent will enable the uranium producer to recover his sunk costs, pay income taxes, and make a reasonable profit. Note, however, that if the average ore grade was found instead to be only 0.14 percent, the ore body would still have been considered by ERDA as part of the ore reserve, but the company mining the deposit would not make a profit, would not be liable for income taxes, and could not fully recover its sunk costs.

In such a situation, a company could continue short-term operations because it could cover its current liabilities, i.e., its operating expenses. However, it could not remain in business indefinitely on such a basis. In spite of this, there is nothing inherently "wrong" with the ERDA approach, particularly since it conveys useful economic information. Such information is more descriptive, and probably more accurate, than simply calculating the estimated ore reserve by grade increments.

A table listing estimated ore reserves by grade would be misleading because some of the high grade deposits are probably unrecoverable due to depth, size (too small), lack of consolidation, remote location, etc. However, if ERDA "marginal cost" ore reserve estimates are used, planners must be aware of the difference between a "forward cost" and a "fair price" for U_3O_8. The former describes ore reserves on a short-term, marginal operating cost basis; the latter should reflect the full cost of resource production over the long term.

A recent prediction of the price of uranium, based on the assumption that the highest production cost incurred (to produce the "last" pound needed) sets the price trend, is that U_3O_8 prices could be about $77 per pound (in 1976 dollars) by 1985, and will likely be no lower than $54 per pound (Davis, 1977, p. 92).

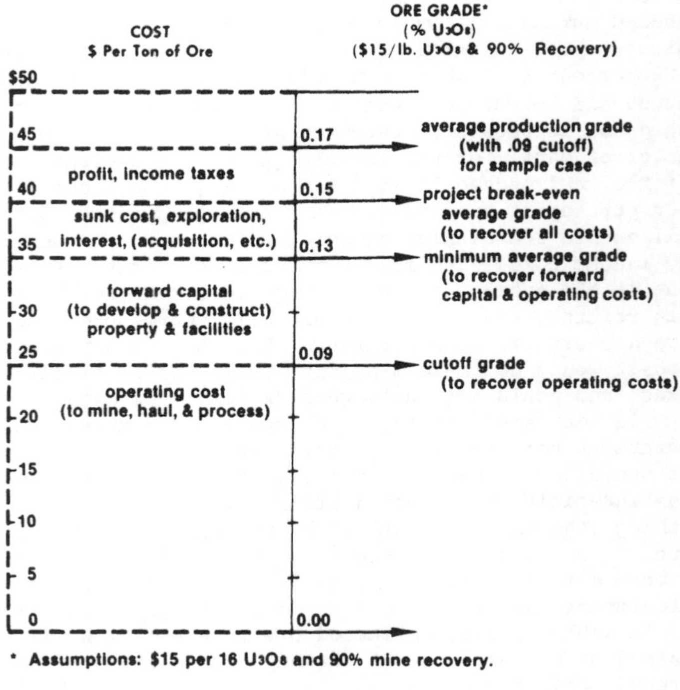

Figure 23.4 Ore Deposit Costs and Ore Grades (developed ore deposits)
Source: Patterson, 1976

COST ESTIMATES

Three principal types of expenditures are of concern to most uranium producers: exploration costs (to find and define a uranium deposit); mining costs (to extract the uranium from the ground); and milling costs (to process the ore into a readily-salable commodity). In Colorado, however, the last type is often the least important. This is because there is a large number of small mines, producing only a few hundred tons of uranium ore per month, that neither own, nor could these small mines afford to build a mill. Instead, these small producers ship their ore to a nearby mill, usually owned by a large producer, or to an ore buying station. In either case, the producer is paid an amount based upon the U_3O_8 content of the ore, without having to mill the ore himself.

In the uranium industry, as in most of the mineral industry, costs have increased rapidly over the past few years due to inflation. However, two other circumstances have also contributed to increasing uranium production costs: a continuing decrease in the grade of uranium ore mined and processed, and a tendency towards greater operating depths for both underground and open-pit mines. Average ore grade produced from open pit mines dropped from 0.16 percent U_3O_8 in 1975 to about 0.12 percent in 1976; the decrease for underground mines was from 0.19 percent to 0.18 percent (Facer, 1976, p. 191). Figure 23.5 indicates the historical average ore grade for both types of mines (Koch, 1977, Fig. 2).

The principal effect of this decline in ore grades is to increase the quantity of ore that must be processed in order to produce a given amount of U_3O_8. Thus, since the mining cost per ton of ore extracted is essentially independent of the grade of the ore, the total amount to be mined and the total mining cost will increase. Similarly, declining ore grades will require larger capacity mills, with a commensurately larger investment for mill construction. Fortunately, the mill operating cost per ton of ore should decrease due to economies of scale. Mill recovery rate will also decrease, from about 95 percent for 0.16 percent ore, to approximately 87.5 percent for 0.05 percent ore. This alone will increase uranium mill throughput requirements by about 10 percent. One beneficial aspect of the declining ore grade problem is that it will allow the extension of existing ore reserves and the development of deposits which were previously uneconomic.

Increasing mine depth is not yet an acute problem in Colorado. However, in New Mexico, the major underground mines will all be 2,000 feet or deeper, with the deepest mine operating level slated at a depth between 3,500 and 4,200 feet (Koch, 1977, p. 7). Such operating depths not only greatly increase operating costs, but also increase the lead time required before a new level can be put into production.

The principal difficulty with estimating costs that would be incurred in developing a particular uranium property is that

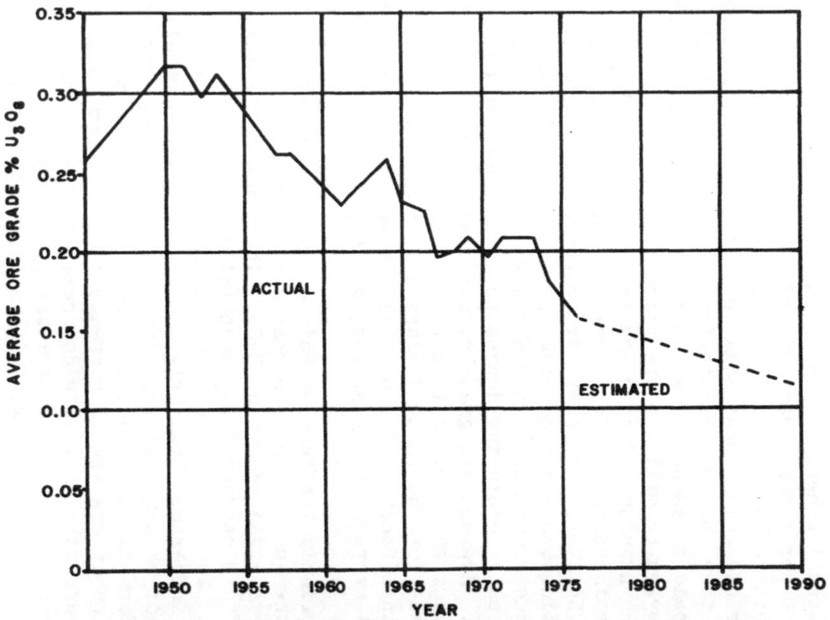

Figure 23.5 Average Mined Ore Grades, U.S. Uranium Mines
Source: Koch, 1977.

each mine and/or mill complex is virtually unique. Factors contributing to this individuality include: physical and chemical characteristics of the ore body (depth, shape, tonnage, grade, and associated minerals); terrain and local environmental conditions; mining method; method and distance of ore haulage; amount and type of equipment and operating supplies to be used; and availability of skilled labor (as well as housing and municipal services available in the area). Additionally, operators may use significantly different approaches in operating the mine and/or mill, simply because of differences in management philosophy or market strategy. While general cost estimates can be developed on the basis of statistics or empirical data, these should be considered as simply "ball park" figures when applied to a particular uranium property. Unfortunately, since most uranium producers are unwilling to divulge their actual costs, costs given in the next two sections are based on general cost data.

Exploration Costs

Table 23.4 lists the typical costs associated with a general uranium exploration program. However, the cost of the actual exploration budget would vary with each particular company, e.g., the type and level of each activity would depend upon its particular goals and resources. Furthermore, any forecast of the total cost necessary to find a uranium deposit is highly speculative because discovery is not guaranteed. The first or the five-hundredth drill hole might indicate an economic uranium deposit, or only "dry" holes may result. As a guideline ERDA estimates the expected time required to make a discovery as approximately five years after start of the search. Thus, according to one source, ". . . for a company to be competitive in uranium exploration, they (sic) should be prepared to spend $1,000,000 to $4,000,000 annually for 5 to 12 years" (Sprouse, 1976, p. 96).

The relationship between exploration costs and total costs in the uranium industry can be assessed by examining the ratio of total exploration expenditure in a given year to total uranium production in the same year. In the words of Mr. Ludwig Koch, Manager, Commercial Development, Continental Oil Corporation:

> We then see that the industry spent about $2 in exploration for every pound of uranium produced in the years 1966 through 1974, increasing to about $6 in 1975 and close to $7 in 1976 . . . this expenditure for exploration is presently equivalent to between 40 - 50 percent of the gross realization from all uranium sales. This compares with maybe 10 percent for the oil and gas industry and not more than 2 or 3 percent for the coal industry. Surely, this disproportionate cash flow situation cannot go on forever in the uranium industry . . .

TABLE 23.4

COSTS INCURRED IN AN EXPLORATION PROGRAM

Land Acquisition	Purchase Cost		
Lease Acquisition	$1.00-$2.00/acre/year		
Claim Purchases	$50-$100/claim/year		
Exploration			
Geological	Cost/Month/Employee		
Professional man months - Permanent	$3500		
Professional man months - Field Agent/Asst	$2500		
Professional man months - Part time	$1850		
Professional man months - Contract Labor	$1100		
Geophysical			
Airborne-Radiometrics	$10-$40/linear mile		
-Magnetics	$10-$15/linear mile		
Ground-Radiometrics	$20-$40/linear mile		
-Magnetics	$20-$40/linear mile		
Three Dimensional Sampling	Low	Avg	High
Core Drilling - 0 to 500 ft	$10/ft	$16/ft	$29/ft
>500 ft	$11/ft	$17/ft	$30/ft
Non-core Drilling - 0 to 2000 ft	$.50/ft	$2/ft	$5/ft
Geophysical Logging Gamma, Resistivity, SP	$.60-.25/ft + transportation over 50 miles		
Contract by month	$2500 + 2 cents/ft		
Contract by week	$ 850 + 2 cents/ft		
Contract by day	$ 10/hole + 10 cents/ft		
Geochemical			
Field Agent	$1600/month/employee		
Sample Analyses	$3-$25/sample		
Other			
Aerial Photography	$10-$15/line mile		
Aerial Geologic Reconnaissance	$5-$10/line mile		
Radon-in soil gas survey			
Instrument measurement of	$20-$25/hole		
Track Etch Cup	$25/hole		

SOURCE: Sprouse, 1976, p. 97.

Table 23.4 (continued)

#1. For a company to increase its chance of successful discovery it would have several exploration projects operating simultaneously.

#2. It is felt by those involved in the search for uranium today that for a company to be competitive in uranium exploration they should be prepared to spend $1,000,000 to $4,000,000 annually for 5 to 12 years.

Mining and Milling Costs

While it is difficult to obtain specific cost estimates for a particular uranium property, there are some general cost estimates available. These cost estimates apply to two types of expenditures: capital investment and operating expenses. The capital investment category includes costs incurred that enable a mine/mill complex to begin production and processing, such as: field expenses, land acquisition, exploration, development, mine facilities and equipment, and mill construction. Operating costs are simply expenses incurred while operating the mine/mill complex. These include: royalties; selected taxes (use, severance, and property); on-going reclamation costs; and the labor, utilities, and supplies (fuel, explosives, repair parts, etc.) required to mine, transport, and process the ore.

The capital expenditure required to develop a 1,000-foot deep ore body by underground mining methods is roughly $80 to $120 (1978 dollars) per annual ton of ore recovered. A 2,000-foot deep deposit could require capital investment of as much as $200 per annual ton. The capital cost of a surface mine can equal the capital costs of the underground mine. This situation is due to the extensive pre-production stripping required to expose the ore body. Mill capital costs are presently about $15,000 per daily ton of capacity (1978 dollars). Table 23.5 indicates the typical capital and operating costs for a new conventional 1,000 ton per day (TPD) open-pit mine and mill facility. Basic assumptions included milling 1,000 TPD of ore containing 0.10 percent U_3O_8 with a stripping ratio of 22 cubic yards per pound of U_3O_8. Milling costs were based on a conventional agitated acid leaching, solvent extraction, and precipitation circuit. A twenty-four month construction period was assumed, and the life of the property was estimated at twelve years (Environmental Protection Agency, 1978, pp. 3-50-52).

The economic impact of these capital and operating costs on a proposed uranium operation depends primarily upon the size and grade of the uranium ore deposit. That is, the U_3O_8 ore reserves must be sufficiently large and rich to generate revenues, in excess of the capital and operating costs incurred over the lifetime of the project. A new producer in 1978 must receive $25 to $30 per pound of U_3O_8 to break even. This break-even

TABLE 23.5

ECONOMICS OF CONVENTIONAL MINING AND MILLING

Operating Costs	\$/ton Ore	\$/lb U_3O_8
Strip and Internal Waste	20.00	11.11
Mining	3.80	2.11
Milling	7.00	3.89
General and Administrative	3.00	1.67
Aquifer Restoration	0.80	0.44
Royalty and Severance Taxes	3.60	2.00
Total	\$ 38.20	\$ 21.22

Investment	\$ Million
Mine Mobile Equipment	9.0
Mill and Tailings	15.0
Mine Shops and Electric	2.5
Roads, Site Preparation, etc.	1.0
Total Capital	\$ 27.5
Working Capital	3.0
Pre-stripping	7.0
Infrastructure	4.0
Total Initial Investment	\$ 41.5

SOURCE: Environmental Protection Agency, 1978, p. 3-51.

value excludes the cost of exploring for new reserves (Environmental Protection Agency, 1978, p. 3-50).

Facing serious cost increases, several companies have been conducting tests on nonconventional, economically promising, mining methods, such as an <u>in-situ</u> uranium leaching program. The estimated capital and costs for <u>in-situ</u> leaching of a 0.05 percent U_3O_8 ore body are shown in Table 23.6, based on a production rate of 250,000 pounds of U_3O_8 per year from a 500-foot deep deposit. Overall recovery was estimated at 60 percent, and the productive life of the deposit was assumed at 12 years to match the open-pit model. <u>In-situ</u> leaching appears to be an economically sound method of recovering uranium from low-grade and other uranium deposits that cannot be mined using conventional methods.

The uranium industry spent $258 million on domestic exploration in 1977; planned domestic expenditure in 1978 is $288.5 million. An estimated $325 million was spent for uranium mine development and construction in 1977, with $422 million planned for 1978. Mill construction expenditures in 1977 are estimated at $167 million with $212 million planned for 1978 (U.S. Department of Energy, 1978a, p. 77). It is clear that a large sum of money will be necessary for the expected uranium industry expansion.

Where will this money come from? As noted earlier, the uranium industry's exploration expenditure presently equals between 40 and 50 percent of its gross realization from all uranium sales. Significant external financing from banks or other financial institutions is unlikely, simply because of the high risk and low credit rating generally associated with the industry. However, this situation may improve in the near future, in light of the strengthening market and price for uranium.

Utilities are asking for delivery of uranium through 1990 at a given price. Unfortunately, producers understandably hesitate to make such long-term commitments because of future uncertainties regarding both the cost and price of uranium. Thus, one alternative open to the uranium industry is to encourage utilities using uranium to share the cost of development. Development, <u>not exploration</u>; utilities should not, and probably would rather not, participate in the risk-taking involved in uranium exploration. This approach would reduce the industry's almost total reliance upon internal financing, that is, financing projects from retained earnings or from the sale of capital stock or bonds. Koch (1977, p. 10) states that: ". . . where utility money is required is squarely in the developing of new mines. This is where the big money is needed and where front-end financing, coupled with a commitment for the sale of the uranium product, would go furthest in reducing the uncertainties now blocking the flow of capital."

If buyer requirements for uranium were to grow from about 17,000 tons of U_3O_8 in 1976 to approximately 40,000 tons in 1985, daily mill capacity must increase from approximately 28,000 tons

TABLE 23.6

ECONOMICS OF SOLUTION MINING

Operating Costs	\$/lb U_3O_8
Wells	12.10
Pumps and Piping	2.32
Power, Coring, etc.	0.86
Milling	6.76
General and Administrative	1.40
Reclamation	0.34
Royalty and Taxes	2.00
Total	\$25.78
Investment	**\$ Million**
Mobile Equipment	2.3
Mill and Tailings	6.5
Roads, Site Preparation, etc.	1.0
Total Capital	9.8
Working Capital	1.4
Initial Well Field	2.2
Infrastructure	1.3
Total Investment	\$14.7

SOURCE: Environmental Protection Agency, 1978, p. 3-53.

of ore per day to nearly 100,000 tons per day over the same period, in order to process a sufficient volume of lower grade ores. This increase in mining and milling capacity would require an investment on the order of $6 to $9 billion, exclusive of the cost of replacing existing facilities or exploring for new deposits (Koch, 1977, p. 9). Significant external financial assistance may be required by the uranium industry to achieve these goals. It is doubtful, as well as undesirable, that the Federal government would provide such assistance. However, it does appear to be beneficial, to both parties concerned, for the electric power industry to consider financing development of selected uranium mines and mills in return for an assured supply of uranium at a set price.

24
Environmental Considerations

Environmental concerns associated with the front end of the uranium fuel cycle, particularly mining and milling, fall into two groups. The first is generally similar to those connected with other mineral projects: surface and wildlife habitat disturbances, and degradation of air and water quality. In the case of uranium, however, there is the added complication of radioactivity associated with each step of the extraction and concentration processes. Consequently, besides traditional environmental consideration, effort must be devoted to measuring and controlling the radiation levels to which working personnel are exposed. This holds true for human, and livestock, in close proximity to the uranium operation.

An evaluation of the environmental impact of a proposed uranium mine and/or mill is generally required at both Federal and State levels of government, since most prospective operations are on federal lands (See Regulatory Considerations, Chapter 25). The National Environmental Policy Act of 1970 requires that an Environmental Impact Statement (EIS) be prepared if ". . . major federal action significantly affecting the quality of the human environment . . ." is involved (Root, 1976).

Since Colorado is an "agreement state" (see Regulatory chapter) an Environmental Report (ER) is required and it is similar in nature to an EIS. Such a report (actually a study) includes the following sections: project description; description of the existing environment; proposed operations and their expected impact on the environment; measures that can be taken to lessen expected environmental impacts; unavoidable adverse environmental impacts; relationship between short-term uses and long-term productivity; and irreversible and irretrievable commitments of resources. As an ER is a rather lengthy and comprehensive report, it may require two to three years to gather baseline data, compile and circulate it, and finally gain approval of such a study. As described in the Regulatory chapter, the Colorado Department of Health is the State's lead agency in coordinating the necessary approvals as a basis for issuing the necessary licenses and permits required before beginning production.

GENERAL RADIATION HAZARDS

Radionuclides (uranium, radium, and their associated "daughter" products (see Figure 24.1) can be absorbed by inhalation or by ingestion as indicated in the pathways shown in Figure 24.1. The principal health hazard of these radionuclides is that they emit ionizing radiation--radiation with enough energy to "break" atoms or molecules into charged fragments called ions.

Two major biological effects may occur when important molecules in the body (such as DNA) are split by ionizing radiation. These are: (1) somatic effects, which affect the normal body functions (such as by increasing the probability of causing cancer); and (2) genetic effects, caused by the mutation of genes or chromosomes (which may adversely affect a person's descendants). It is important to note that human genetic changes have yet to be observed, even when massive radiation doses have been absorbed (Beckmann, 1976, p. 59). Several factors influence the biological effect resulting from exposure to radiation: the amount of the radiation absorbed; the rate at which this radiation is delivered (dose rate); the size (and location) of the body area exposed; and the individual characteristics of the person receiving the radiation.

In this regard, it is necessary to note that the existing environment subjects man to constant exposure from natural radiation sources. These uncontrollable exposures result from radioactive materials in the earth's crust (uranium, thorium, and potassium), radionuclides in air and water and cosmic radiation (high energy charged and uncharged nuclear particles from stellar sources continually entering the earth's atmosphere). The cosmic radiation effectively doubles for each one mile increase in altitude above sea level.

Additionally, the radiation background contributing to man's exposure is augmented by man's activities. Sources of radioactivity include medical treatment and diagnosis, nuclear weapons testing, burning of fossil fuels, high altitude air travel, fertilizer use and proximity to mining and milling activities. This chapter examines only the latter, but it is noteworthy that where and how one chooses to live are important determinants of equantities of low-level radiation exposure (See Table 24.1).

Mining and processing uranium can increase radiation exposure to human populations and ecosystems in the vicinity of these operations, if not controlled. In uranium mining and milling only about 10 - 15 percent of the radioactive material in the raw ore is removed; some 85 - 90 percent remains in mill tailings (Environmental Protection Agency, 1978, p. 4-39). Thus, processing or milling presents the largest single environmental concern.

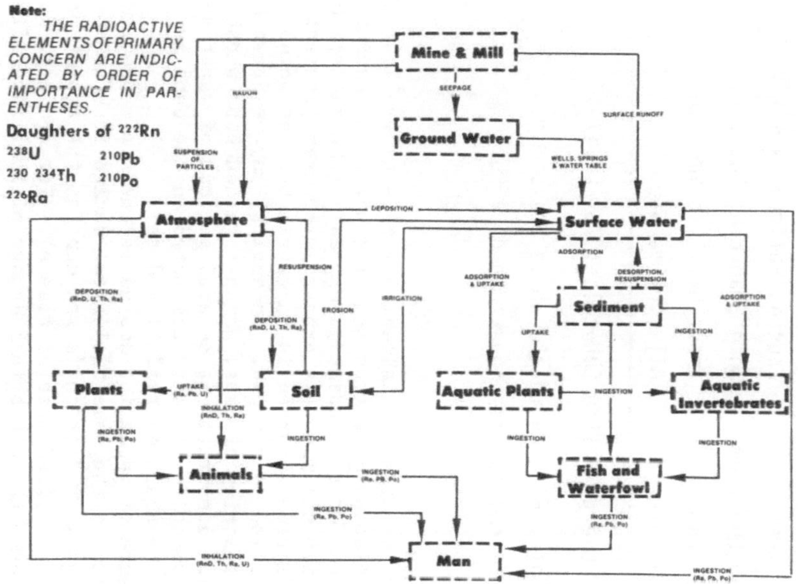

Figure 24.1 Transport and Movement of Radionuclides to Man
Source: Whicker and Johnson, 1978.

TABLE 24.1

SHORT TERM RADIATION DOSAGE EFFECTS

Radiation Dose Received*	Effects Evident Within a Few Days
Under 25,000 mrem**	No Observable Effect
About 50,000 mrem	Slight, Temporary Blood Changes
About 100,000 mrem	Nausea, Fatigue, Vomiting
200,000 to 250,000 mrem	Some Chance of Fatality
About 500,000 mrem	About Half Victims Die
About 1,000,000 mrem	All Victims Die

SOURCE: Iowa Energy Policy Council, 1975, p. 17.

*Single dose of radiation to the whole body.

**A millirem (mrem) is one-thousandth of a rem.

Radiation records maintained on miners generally reflect either the radiation dose (the actual quantity of radiation energy absorbed) or the radiation exposure (the radiation intensity to which the miner is exposed); however, the actual amount of radiation absorbed is uncertain because actual inhalation/retention is unknown. An analogy may be drawn here with filling a series of cups, using a garden hose whose flow rate (radiation dose rate) is several gallons of water per minute. The quantity of water collected in the cups is analagous to the absorbed radiation dose.

One measure of radiation dosage is the rem. One rem of radiation produces the same biological damage in a human as 0.97 rads of gamma radiation (a rad denotes the amount of energy absorbed by a specified mass of tissue). The rem unit is necessary because the biological damage of ionizing radiation is not directly proportional to the amount of energy absorbed; that is, one rad of alpha radiation is more damaging than one rad of gamma radiation.

The rem or rad represents a rather large quantity of radiation, hence the term millirem (mrem or one thousandth of a rem) is more commonly used to measure radiation dosage. The International Commission on Radiological Protection, under the aegis of the United Nations, has set the value of 500 mrems as the maximum permissible annual dose that an individual should receive. The figure is conservative, since there are areas in India and Brazil in which monazite sands, rich in thorium and uranium, expose the population to an average annual dose of 1,500 mrems (Beckmann, 1976, p. 56).

Table 24.1 gives the short-term effects of a single dose of radiation on the whole body. However, the long-term effects of continued exposure to low levels of radiation are not fully known.

The natural background radiation in the United States amounts to an average whole-body dose of about 102 mrem per year; in the Denver, Colorado, area the natural background dose is approximately 154 mrem per year, as shown in Table 24.2.

The Advisory Committee on the Biological Effects of Ionizing Radiation of the National Academy of Sciences estimates that the natural background radiation of 102 mrem could cause 1,800 to 9,000 cancer deaths annually in the United States (Iowa Energy Policy Council, 1975).

Allowable radiation doses have been lowered several times in the past twenty years. Currently, artificial radiation doses (those from medical uses, fallout, and industrial environments) are limited to an average of 170 mrem per year. Highly variable, but approximately average figures for artificial radiation sources are: 70 mrem/year from x-rays and radioactive materials used in medical diagnosis and therapy, and 4 mrem/year for fallout from atmospheric nuclear weapons tests. The National Academy of Sciences Advisory Committee estimates that exposure to an average of 170 mrem/year, over a 30-year period, would increase the spontaneous cancer death rate perhaps 1 to 5 percent. This would mean an additional 3,000 to 15,000 cancer deaths annually in the United States (Iowa Energy Policy Council, 1975).

The usual measure of exposure, as opposed to the dose, is the "Working Level" (WL) radiation, expressed in terms of the potential alpha energy present (in picocuries, see Glossary) per liter of air. The terms "Working Level Hours" and "Working Level Months" consequently describe cumulative exposure, based on intensity and duration, to radionuclides, rather than the actual dosage absorbed into the lungs. Colorado has adopted an EPA radiation exposure guideline established in 1970: maximum exposure permissible in one calendar year is four working-level months. While exposure to one working-level month is not strictly comparable to dosage, due to uncertainty of inhalation, retention, absorption, etc., an approximate correspondence can be indicated. That is, <u>if</u> one working-level month's exposure was actually inhaled, assuming average respiration rates, retention rates, etc., then this would be approximately equivalent to a 6 rem dose to the lungs (National Institute on Occupational Safety and Health, 1971). The Radon-Daughter Control Program of the Colorado Division of Mines resulted in the taking of 1.047 radon-daughter samples in 313 inspections at 84 mines in 1976. Table 24.3 shows intensity distribution for the radon-daughter samples taken (Colorado Division of Mines, 1976).

TABLE 24.2

ESTIMATES OF AVERAGE ANNUAL WHOLE-BODY DOSE RATES IN U.S. (1970)

Source	U.S. Average Dose Rate (mrem/yr)	Colorado (mrem/yr)
Environmental		
1. Natural		
Cosmic Radiation	44	90
Terrestrial Sources[(a)]	40	46
Internal Sources[(b)]	18	18
Total	102	154
2. Global Fallout	4	4
3. Nuclear Power	0.003	0.003
Medical		
1. Diagnostic[(c)]	72	72
2. Radiopharmaceutical	1	1
Occupational	0.8	0.8
Miscellaneous	2	2
Totals	182	234

SOURCE: Iowa Energy Policy Council, 1975, p. 17; Whicker and Johnson, 1978.

(a) Crustal and building materials
(b) Principally potassium - 40
(c) Based on abdominal dose

TABLE 24.3

DISTRIBUTION OF RADON DAUGHTER SAMPLES* (1.0 WL represents 300 picocuries per liter of air)

Range	Number of Samples	Percent	Cumulative Percent
0 - 0.3 WL	626	59.79	59.79
+0.3 - 1.0 WL	339	32.38	92.17
+1.0 - 2.0 WL	42	4.01	96.18
+2.0 - 5.0 WL	30	2.86	99.04
+5.0 - 10.0 WL	5	0.48	99.52
+10.0 WL and over	5	0.48	100.00
Totals	1,047	100.00	

SOURCE: Colorado Division of Mines, 1976.

*Collected in Colorado mines.

CONVENTIONAL MINING

In 1977, 92 percent of United States uranium ore was produced from conventional mining operations, that is, underground or open-pit mines. In Colorado, virtually all production is from underground mines, with only one major open-pit mine in the Marshall Pass area of Saguache county proposed. With only a few exceptions, most underground mines in Colorado are small operations, employing only three or four people and producing only a few hundred tons of ore per month (Appendix E).

Major environmental impact categories, aside from radiation hazards, are essentially the same as for coal operations: surface disturbance and degradation of air and water quality. However, uranium mining generally requires the removal of much smaller quantities of material than coal mining, due to the higher energy content of uranium ore, regardless of whether the mine is underground or open-pit. Coal mines may produce several million tons of coal per year, whereas uranium mines usually only produce a maximum of several hundred thousand tons of ore. Also, the disturbed surface area is significantly less for an underground uranium mine than for an open-pit uranium mine. Surface requirements for small underground uranium mines operating in the Uravan Mineral Belt have been estimated as follows: about 1/10th acre for the underground mine portal; about 1 acre for surface storage of waste rock (removed from the mine shaft, etc.); from 1 to 5 acres for surface facilities, such as ore bins, hoist house, change house, mine office, parking area, and maintenance facilities; and approximately 2 acres per section of land for drill roads and access roads, assuming a 10-foot roadway (U.S. Atomic Energy Commission, 1972). While these numbers appear somewhat conservative, it was the most up-to-date data on Colorado's underground uranium mines.

While it is generally too costly to return waste rock to underground "rooms," in most cases waste dumps can be graded to blend with existing land. Depending upon the particular lease or permit stipulations, such dumps may have to be covered with topsoil and seeded, with contoured drainage to prevent contamination of any water runoff. Most uranium mines in Colorado are located in semi-arid areas, with the uranium ore body well above the water table. However, if mine dewatering (and subsequent discharge to the local area) is necessary, the operator is required to prevent surface runoff from entering (and being contaminated by) the mine workings.

Underground mining activities also add a minor amount of pollutants to the air outside the mine, principally through dust and carbon monoxide from engine emissions. However, the greatest air quality problem is inside the mine--generation of radon gas and its daughters, by uranium ore. Special ventilation systems are required to reduce and maintain airborne radioactivity to acceptable levels. Large-capacity, air circulation pumps are used with special exhaust shafts to provide adequate ventilation. The two basic ventilation guidelines are: (1) channel the fresh air supply so that it displaces, but doesn't mix with, contaminated air; and (2) seal off old (exhausted) working areas so that additional radon is not introduced into the current working areas. Radon gas in the mine ventilation exhaust is usually of very low concentration and is quickly and readily dissipated into the outside atmosphere. Previous tests have shown exhausted radon gas is of little significance off the mine site (U.S. Atomic Energy Commission, 1972). However, the EPA reports some concern regarding radon gas emanation from mine vent holes in the Grants Mineral Belt, especially since houses are nearby (McGinley, 1977). Finally, while no data are available on land subsidence from abandoned underground mines, such subsidence should be considerably less than that from either underground coal or oil shale operations, because of the smaller volume of material extracted (University of Oklahoma, 1975).

The environmental impacts of an open-pit uranium mine differ from those of an underground mine. Here, any radon gas emissions are readily vented to the atmosphere, and are expected to be undetectable. However the total surface area disturbed can become significant, due to the amount of overburden that must be removed. Table 24.4 lists the environmental residuals associated with a hypothetical open-pit mine. The data have been normalized to fit the case of an open-pit mine supplying the U_3O_8 required by a 1,000 Megawatt (MW) light water reactor (LWR).

TABLE 24.4

SUMMARY OF ENVIRONMENT RESIDUALS FOR URANIUM MINING
(Normalized to 1,000 MWe LWT Annual Fuel Requirement)

Natural Resource Use	Quantity/Year
Land (acres)	
Temporarily committed	55
Undisturbed area	38
Disturbed area	17
Permanently committed	2
Overburden moved (millions of metric tons)	2.7
Water (millions of gallons)	
Discharged to ground	123
Effluents	
Chemical (metric tons)	
Gases*	
Sulfur oxides	8.5
Nitrogen oxides	5.0
Hydrocarbons	0.3
Carbon monoxide	0.02

SOURCE: University of Oklahoma, 1975, p. 6-11.

*Estimated effluent gases based upon combustion of coal to supply power; together with combustion of diesel fuel for mining equipment operation.

Assuming an average ore grade of 0.20 percent, this open-pit mine would produce approximately 100,000 tons of ore, yielding approximately 200 tons of yellowcake (U_3O_8) annually. As previously stated, most uranium mining areas in Colorado are semi-arid. Consequently, it is extremely doubtful that an open-pit uranium mine in Colorado would require mine de-watering anywhere near the amount given in the table. Siting factors for conventional mining and milling are listed in Table 24.5.

SOLUTION MINING

When compared to conventional mining systems, solution mining of uranium minimizes surface disturbance and aerial discharge of radionuclides, as shown in Table 22.3. (See Uranium Processing, Chapter 22.) Surface disturbance is generally limited to access roads and surface facilities. Air quality

TABLE 24.5

FACTORS AFFECTING MINE AND MILLING SITING

1. Proximity to mine	9. Cultural features
2. Topography	10. Land use - mine development and operation (mine de-watering)
3. Proximity to residents, agriculture and animals	- overburden and liquid waste disposal
4. Geology, geochemistry and hydrology	- ore stockpiling
5. Soils and overburden	- access roads for ore haulage
6. Meteorology	- utility corridors
7. Biology - food chain transport	
8. Seismicity	

SOURCE: U.S. Environmental Protection Agency, 1978.

degradation is generally limited to dust and exhaust emissions from operation of surface equipment. Almost no alpha or beta radiation is vented to the atmosphere since the selectivity of the leaching process leaves the insoluble radon-daughter elements in the ground (Hunkin, 1975). Finally, solution mining interferes with natural groundwater quality and distribution only minimally. Hunkin states that:

> . . . the current practice of using alkaline leach solution permits 100 percent recirculation, so that the volume of fluid in the operating formulation remains constant, and the natural hydraulic gradients are unchanged outside the operating zone. As a result, solutions are confined to the area under treatment. . . .The very dilute leach solutions, compatible with natural groundwaters, ensure that no environmental damage results, even if a loss of control occurred.

Solution mining of uranium appears to offer significant environmental advantages over underground and open-pit mining. Surface disturbance is minimal and its reclamation is relatively simple. Further, there are no mill tailing piles resulting from this mining technique; the leachants used are selective for uranium, leaving its strongly radioactive daughters within the residue of the subsurface ore body. Nonetheless, environmental concerns exist over being able to recover all the leachants in order to eliminate any possibility of groundwater contamination. Site-specific testing and demonstration are generally required before issuing a mining permit.

MILLING

A uranium mill extracts and concentrates more than 90 percent uranium oxide (U_3O_8) from uranium ore by a series of mechanical and chemical processes. (See Uranium Processing, Chapter 22.) The liquid and solid wastes from this milling process contain radioactive decay products (fourteen radionuclides) principally radium-226; airborne wastes include radon gas (the decay product of radium), its daughter products, and fine ore particles. The solid mill wastes, or tailings, are of special interest since about 90 percent of the original ore is converted to waste. Disposal of these leached solid wastes is a problem which must be properly addressed. Table 24.6 lists the environmental residuals associated with a mill annually supporting a 1,000 MW LWR, e.g., a mill processing approximately 100,000 metric tons of ore annually. Liquids must be caught in a tailings pond, or they may be treated to reduce the concentration of radioactive and other undesirable elements to acceptable levels before being released. Otherwise, waterborne radium-226 could be concentrated by alfalfa, hay, algae, or aquatic plants, and later pose a health hazard. The solid tailings must be permanently impounded to preclude escape of harmful substances to the surrounding environment by wind erosion, surface water runoff, or leaching out by rainfall. The Colorado Department of Health developed regulations for the stabilization of mill tailing piles to minimize wind and water erosion and the escape of radionuclides. These regulations became effective in January 1967 (Beverly, 1977).

The control of radioactivity released from mill tailings is covered by federal regulations which have been adopted by agreement states such as Colorado. (See Regulatory Considerations, Chapter 25.) Underground disposal is recommended by the Nuclear Regulatory Commission (NRC). If not feasible, tailing piles must be stabilized under several feet of clay plus topsoil for vegetation. The clay cap is used to reduce radon emanation from tailing (radon is produced at a daily rate of 0.18 Ci radon per Curie (Ci) or radium) (Environmental Protection Agency, 1978, p. 4A-3), to no more than twice that naturally produced in the local area.

Additionally, a monitoring program is required during the lifetime of the project as well as for a defined period following the end of milling. It is important to note that the maximum opportunity for radon release (if controlled) occurs at the end of the milling operation, i.e., the existence of maximum quantities of tailings--and maximum amount of radium present.

The components of a sampling and monitoring program currently performed at Grants, N.M., by the United Nuclear Corporation, are listed in Table 24.7. Concerns are reflected in

TABLE 24.6

SUMMARY OF ENVIRONMENTAL RESIDUALS FOR URANIUM MILLING (Normalized to 1,000-MWe LWR Annual Fuel Requirement)

Natural Resource Use	Quantity/year
Land (acres)	
Temporarily committed	0.5
Undisturbed area*	0.2
Disturbed area	0.3
Permanently committed (limited use)	2.4
Water (millions of gallons)	
Discharged to air	65
Effluents	
Chemical	
Gases** (metric tons)	
Sulfur oxides	37
Nitrogen oxides (40 percent from natural gas use)	15.9
Hydrocarbons	1.3
Carbon monoxide	0.3
Liquids (thousands of metric tons)	240
Tailings solutions	
Solids (thousands of metric tons)	91
Tailings solutions	
Radiological (curies)	
Gases (including airborne particulates)	
Radon-222	74.5
Radium-226	0.02
Thorium-230	0.02
U natural	0.03
Liquids	
U and daughters	2
Solids	
U and daughters	600
Thermal (billions of Btus)	69

SOURCE: University of Oklahoma, 1975.

*Major portion of undisturbed area for mills is included in mine land use.
**Estimated effluent gases based upon combustion of equivalent coal and natural gas for power and heat.

TABLE 24.7

SAMPLING AND MONITORING PROGRAM

Medium	Parameters	Frequency
Surface Water	Flow Rate Temp.,	Continuous
	Total Soluble Solids	Weekly
	U, Ra, Mo, Se, V	Monthly
	Alpha, beta, U, Th, Ra, Chem. Series	Semiannually
Ground Water	Alpha, beta, U, Th, Ra, Chem. Series	Semiannually
Air Particulates	Alpha, beta, U, Th, Ra	Semiannually
Breathing Air in Mill	Alpha, beta gamma	Monthly
Soil	Alpha, beta, U, Th, Ra	Semiannually
Vegetation	Alpha, beta, U, Th, Ra	Semiannually
Dose Rates	Gamma	Semiannually
Tailings Pond Water	pH, U, Th, Ra	Operationally Determined
Meteorology	Wind speed & direction, temp., humidity, PPT	Continuous

SOURCE: Sauvingac, 1978, p. 132.

measuring radioactivity and in air and water quality. The practice of using some of the sand tailings from a uranium mill site as fill in construction projects was halted in 1966. Operations are still underway to correct the problems created by this practice. Corrective action includes removing this contaminated fill and replacing it with "clean" fill. Presently, written approval of the Colorado Department of Health is required before any tailings can be removed from any active or inactive mill in Colorado. The potential hazard of radon emanation from tailing piles, from low concentrations over long periods of time, is also being investigated. The maximum mill radiation exposure guideline will be a total of 25 mrem per year from each individual nuclear fuel cycle operation beginning in 1980 (Beverly, 1977).

25
Regulatory Considerations

Most laws and regulations governing the manufacture, transport, use, and disposal of radioactive materials have been in existence for some time. Between the end of World War II and early 1975, the Atomic Energy Commission (AEC) had the responsibility to collect and evaluate the large body of existing data on radiation effects on health and the environment, and it is estimated that more than $1 billion was used for this purpose. When the AEC was dismantled in January 1975, its research arm became the nucleus of ERDA (now DOE), and its regulatory functions were taken over by the Nuclear Regulatory Commission (NRC).

In 1970, the Environmental Protection Agency (EPA) was given the authority to regulate and monitor radiation in certain circumstances, such as setting limits for radioactivity in air and water quality, and radiation exposure standards.

Since the late 1950s the states have greatly increased their responsibilities for enforcing and monitoring the federal standards for measuring and mitigating environment effects of nuclear development occurring within their borders. Congress enacted Section 274 of the Atomic Energy Act of 1954 to recognize states' interest and involvement in nuclear activities, to clarify the division of responsibilities between the states and the Federal government, and to enable their formal agreement or licensing authority with the AEC (now NRC) over source, byproduct, and special nuclear (fissionable) materials (Environmental Protection Agency, 1978, Section 4).

Those states which have been delegated licensing authority are called agreement states. As of May 1978, there were twenty-five agreement states, of which Colorado is one; each has its own licensing programs, generally embodying all of the federal standards.

Each facet of the uranium industry is subject to a number of regulations and rules, most of which center on environmental and health issues. Any uranium mining or milling facility is subject to federal and State regulations concerning air pollution, water pollution, solid waste disposal, land use, and reclamation requirements in addition to regulations covering exposure to radiation.

The major regulatory agencies which oversee uranium facilities include the Environmental Protection Agency (EPA) and the Nuclear Regulatory Commission (NRC) at the federal level, and the Colorado Department of Health, and the Colorado Department of Natural Resources at the State level. Licensing and enforcement of regulations are carried out at the State level (unless unusual circumstances lead to federal intervention). Since Colorado is an agreement state, it has taken on the responsibility to enact legislation meeting federal guidelines and to enforce these regulations.

The State Mined Land Reclamation Board (within the Department of Natural Resources) oversees land use and reclamation permits; the Division of Mines (also within the Department of Natural Resources) oversees mine and/or mill operating permits; and the Water Quality Control Division, Air Pollution Control Division, and Radiation and Hazardous Wastes Control Division (all within the Department of Health) oversee water pollution, air pollution, and radiation control regulations, respectively. It is this last category, that of radiation control, which may be most significant for the uranium industry because of the potential health hazards.

The Uranium Mill Licensing Guide (Colorado Department of Health, 1978a) serves as a resource document providing applicants, consultants, and the general public with detailed information required in a license application and environmental report (ER) for the conventional milling of uranium, thorium or other radioactive ores, and residues. This includes radioactive waste materials not associated with reactor fuel or spent reactor fuel processing. The guide is also applicable to heap leaching and <u>in-situ</u> solution mining, since it equates the latter operations with milling. All requirements of the State of Colorado Rules and Regulations Pertaining to Radiation Control (Colorado Department of Health, 1978b) and other State statutes must be met by any proposed or existing uranium facility, and it should be noted that following the Licensing Guide in the preparation of a radioactive material license application an ER does not assure or imply approval of any proposed project. Appendix F lists the references pertaining to the uranium milling licensing process that the licensing applicant should obtain, read, and understand.

Additionally, the EPA in 1977 published its "Environmental Radiation Protection Standards for Nuclear Power Operations," which applies to uranium fuel cycle facilities including milling operations. The standards become effective for uranium mills on December 1, 1980, and establish an annual radiation dose limit to any member of the public. The annual radiation dose resulting from uranium fuel cycle operations is 25 millirems to the whole body, 75 millirems to the thyroid, and 25 millirems to any other body organ. The EPA standard excludes radon and its daughters, and can be exceeded only if the regulatory agency has granted a variance based upon its determination that a temporary and unusual condition exists and continued operation is in the public interest. The limit of twenty-five millirems is substantially less than the current maximum limit established by the NRC. The Colorado Department of Health will be the regulatory agency for implementation of the standard (Martin, 1977, p. 230).

With respect to the front end of the nuclear fuel cycle, following the decision to proceed with a mining and/or milling project, the operator must decide where to site the facilities (the mine is fixed; the mill is not) and how to mine and process ore in the most economic and environmentally safe manner. The latter includes both the workers and the general population in proximity to the workings.

Decisions are influenced by many factors, particularly existing regulations (local, state, and federal) which cover all project phases from pre-mining through the active life of the project to post-reclamation surveillance. As noted above, agreement states follow basic federal guidelines, but the states control the process. Colorado's Department of Health (Division of Radiation and Hazardous Waste Control) is the lead agency in the State. The other agencies are listed in Table 25.1. If the workings are on federal lands, the U.S. Bureau of Land Management, U.S. Forest Service, U.S. Bureau of Reclamation, and Bureau of Indian Affairs may be involved in "rights of way" or special land use applications. Procedures: (U.S. Environmental Protection Agency, 1978, Section 4).

1. During early development of project plans, the operator contacts the appropriate regulatory agencies to coordinate compliance with requirements. These influence the engineering and economic feasibility of the project.

2. The operator works closely with the above agencies to insure completeness of applications and to develop a realistic schedule that reflects the review procedures.

3. This should clarify the number and types of required permits and the supporting information for each, and should finalize the program.

4. The applicant then collects baseline data (including radiological information), provides results of premining investigations, prepares and files applications, and submits detailed project plans.

5. The lead agency circulates the documents to the appropriate agencies for their comments; these are incorporated into the Environmental Statement.

TABLE 25.1

LICENSE APPLICATION AND ENVIRONMENTAL REPORT PROCESS*

1. Colorado Department of Health:
   - A. Water Quality Control Division
   - B. Air Pollution Control Division
   - C. Laboratory
   - D. Public Relations Office
2. Colorado Department of Natural Resources
   - A. Mined Land Reclamation Board
   - B. Geological Survey
   - C. Division of Wildlife
   - D. Division of Water Resources
   - E. Division of Mines
3. Colorado Department of Highways
4. Colorado Department of Local Affairs
   - A. Land Use Commission
5. Office of the Governor
   - A. Socioeconomic Impact Office
6. State Historical Society
   - A. State Archaeologist
7. U.S. Nuclear Regulatory Commission
8. U.S. Environmental Protection Agency

SOURCE: Colorado Department of Health, 1978a.

*This list is subject to change at the discretion of the Colorado Department of Health.

6. The project plans are reviewed by the lead agency to determine if bonding requirements are satisfied; designs have been developed to control and mitigate negative environmental effects; the safety of the public is assured.

7. Public hearings are generally required once the lead agency completes its review process.

8. Following completion of steps 6 and 7, the permit may be granted; granted with stipulation (which often includes establishing performance bonds and/or specific monitoring procedures); or denied.

26
Colorado Uranium: Importance and Impact

Mineral resource production has long been a major source of income in Colorado. Table 26.1 summarizes Colorado mineral production for the years 1975 - 1977. Historically, production of uranium and vanadium has only been a small fraction of the total value of Colorado's mineral production. In 1977, the value of the uranium and vanadium produced was only 14 percent of total metallic mineral production and only 5 percent of total State mineral production. However, these figures understate the importance of uranium and vanadium to the economies of several rural areas in Colorado.

The 1977 uranium and vanadium production in Colorado came from seven counties, as indicated in Table 26.2. Six of these counties are in western Colorado, and contain less than 10 percent of the total State population. The value of the uranium and vanadium produced (Table 26.1) reflects only the average value received, at the mine, for unprocessed ore. The value added by milling and concentrating is significantly greater. In fact, the Union Carbide Corporation estimates that its total expenditures for payroll, benefits, taxes, supplies, and services related to uranium and vanadium production in Colorado in 1976 amounted to approximately $45.5 million (Union Carbide, 1977). Such expenditures, whether paid directly (as wages) or indirectly to a contractor or supplier, help encourage and support other local industries, such as wholesale and retail trade, transportation, and consumer services. Consequently, while the uranium and vanadium production is only a small percentage of Colorado's total mineral production, it probably contributes significantly to the economic health of the rural counties of western Colorado.

EMPLOYMENT AND MANPOWER ESTIMATES

In contrast to existing molybdenum operations and proposed oil shale operations, most uranium operations in Colorado are rather small. Out of a total of 130 uranium mines either in

TABLE 26.1

PRODUCTION OF METALS, NONMETALLICS, MINERAL FUELS
(Totals for Colorado)

Metallics	1975	1976	1977
Molybdenum	$146,636,557	$183,448,084	$276,538,944
Uranium	12,585,840	12,984,077	33,411,581
Vanadium	9,835,382	6,965,664	25,041,601
Zinc	28,313,631	24,559,172	20,292,427
Silver	12,512,045	14,292,695	18,560,061
Tungsten	10,243,190	12,868,738	15,544,678
Lead	9,152,549	9,326,657	10,181,349
Gold	10,112,265	5,459,111	8,528,605
Copper	4,198,791	2,983,230	2,001,737
Tin	611,849	593,124	749,306
Cadmium	618,048	425,513	387,771
Iron	692,087	373,494	309,764
Miscellaneous Metallics	9,800	25,441	---
Total Metallic Mineral Production	$245,522,034	$274,305,000	$411,547,824
Total Nonmetallic Mineral Production	$ 76,226,199	$ 75,455,544	$ 83,145,583
Total Mineral Fuel Production	$498,081,903	$632,120,104	$704,641,196
TOTAL STATE MINERAL PRODUCTION	$819,830,136	$981,889,648	$1,199,334,609

SOURCE: Colorado Division of Mines, 1978, p. 9.

TABLE 26.2

COLORADO URANIUM AND VANADIUM PRODUCTION IN 1977 DOLLAR VALUE OF METAL PRODUCTION*

County	Uranium	Vanadium	Total Uranium/ Vanadium
Jefferson	$ 2,034,565	$ ---	$ 2,034,565
Mesa	2,187,410	950,106	3,137,516
Moffat	1,050,440	---	1,050,440
Montrose	20,200,488	16,640,590	36,841,078
Rio Blanco	116,729	16,877	133,606
Saquache	174,475	---	174,475
San Miguel	7,647,474	7,434,028	15,081,502
Total	$33,411,581	$25,041,601	$58,453,182

SOURCE: Colorado Division of Mines, 1978, pp. 10-24.

*Based on the following average prices received:

Uranium $18.36 per pound

Vanadium $ 2.04 per pound

TABLE 26.3

ESTIMATED TOTAL URANIUM EMPLOYEES* IN COLORADO, 1977 (Mine Development and Production)

County	Underground	Surface	Mill	Exploration	Total
Boulder	3	1	3	-	7
Fremont	-	-	99	-	99
Garfield	2	1	41	-	44
Jefferson	100	27	-	-	127
Mesa	87	14	-	-	101
Moffat	-	9**	-	-	9
Montrose	226	42	180	-	448
Rio Blanco	2	2	-	-	4
Saguache	-	9***	-	-	9
San Miguel	204	79	-	-	283
Teller	-	-	-	12	12
Weld	-	-	-	15	15
Total	624	184	323	27	1,158*

SOURCE: Colorado Division of Mines, 1976-1977, Inspection Sheets.

*Approximate numbers only; according to the Colorado Division of Mines, the total number of employees in the uranium industry in 1976 in Colorado was 1,096.

**Heap leaching operation.

***Homestake Pitch Mine exploration and development.

production or under development, only 32 employ more than 5 workers. (Appendix E is a detailed list of uranium operations identified in Colorado in 1977.) The largest currently-producing mine, the Schwartzwalder in Jefferson County, employs approximately 115 people. Table 26.3 lists by county the estimated number of employees in the uranium industry in Colorado; however these numbers are only approximate, since they probably do not provide a complete picture of the actual number of persons exploring for uranium in Colorado. While only 27 persons were identified from mine inspection sheets, the actual number exploring for uranium in Colorado is probably in excess of 200. (Estimated numbers of uranium explorers in Colorado in past years are: 167 in 1973; 205 in 1974; and 235 in 1975 [U.S. Bureau of Mines, 1975a]). Considering this data, the total number of people employed in the Colorado uranium industry (exploration, development, mining, and milling) is probably close to 1350.*

Uranium mining operations in Colorado have been increasing at a rate of about 8-9 percent a year: 110 operations in 1975, 119 in 1976, and 130 in 1977. A leveling-off appears to be occurring in 1978. The current prices and the increasing demand for uranium should continue to fuel this growth, resulting in increased activity in both conventional and unconventional uranium recovery techniques. Ranchers Exploration and Development Corporation recently began leaching operations on old uranium mill tailings in Naturita; the Union Carbide Corporation is using a heap-leach technique in Maybell (Chenoweth, 1977a). Reopening abandoned properties, as well as developing new lower-grade deposits, is a distinct future possibility. What was waste material just a few years ago may be valuable ore today.

Most new Colorado uranium operations will probably be on a small scale if the history of past development is any indicator. However, at least one new mine will be a large open-pit operation, producing about 500-600 tons of ore per day, or approximately 150-200,000 tons per year. This is the Pitch Project, a mine and mill complex announced by Homestake Mining Company in 1976. The manpower requirements estimated for construction and operation of this project are listed in Table 26.4. The proposed long-term mine work force of 136 people would make the Pitch mine the largest single uranium mine in Colorado. In spite of this, the combined mine/mill manpower requirements (206) are still less than one-fifth of the estimated 1,172 person work force required to operate a 50,000 barrel per day underground mine and surface retort oil shale complex (Shell Oil, 1976).

While no large underground mines have been announced, the estimated manpower needed to support a 1,080 tons-per-day underground uranium mine--producing about 300,000 tons per year--is 275 men (Kelley, 1975). In May 1977, Pioneer-Uravan announced

*In 1976, Mr. Chenoweth (DOE) estimated that 375-400 persons were engaged in uranium exploration in Colorado, thus the total number of persons in the Colorado uranium industry may be as high as 1,500.

TABLE 26.4

ESTIMATED MANPOWER REQUIREMENTS: PITCH PROJECT

	1977		1978			1978-1998
	June-Aug	Aug-Nov (peak)	Dec(77)-May	June-Aug	Aug-Nov	(long term operation)
Construction workers	0-86	86	25	25-150	150	0
Mine Operation:						
Pre-production	10-40	40	40	0	0	0
Miners	0	0	0	40-136	136	136
Mill/Office:						
Personnel	0	0	0	0	0	70
Total	10-126	126	65	65-286	286	206

SOURCE: Dames and Moore, 1976.

that it plans to construct, jointly with other companies, a uranium-vanadium mill in either southwestern Colorado or southeastern Utah (Chenoweth, 1977a). No formal manpower requirements are available for such a new mill, although will probably be about the same size as mills currently operating in Colorado. Manpower requirements for these mills fall somewhere between 41 (Rifle mill) and 180 (Uravan mill) employees.

While operations and employment in the uranium industry are expected to increase through the 1980s, two factors tend to discourage boom-type development in Colorado: first, the historic trend toward small uranium mining operations, probably aided by a shortage of skilled uranium mine workers; and second, even large uranium operations require significantly fewer persons, with significantly less impact than other mining operations (such as oil shale, coal, or molybdenum).

SOCIO-ECONOMIC IMPACTS

Construction and operation of a uranium mine and/or mill can provide many benefits to the area in which it is located. These benefits include on-site employment opportunities, encouraging local service companies, and additional tax revenue for local governments. The principal drawback is that if growth is too rapid, more than about 15 percent yearly, the local benefits will only be realized gradually, while negative effects will be immediate.

Negative effects stem from a sudden influx of people--similar to the old-time "rush to the diggings"--that generates a sudden, sharp increase in demand for public and private facilities and services. However, a "boom-town" phenomenon will generally <u>not</u> take place if the existing facilities and services can adequately accommodate the increase in population. Thus, if the population increase is sufficiently small, or if an excess of facilities and service capability is sufficiently large, the negative impacts could be minimal. (Excess services would result from a declining population, as was the case in many areas during the 1960 - 1970 period.)

A long-term undesirable result of excessive annual growth is the tendency of local governments to go into debt to finance crash programs to meet population needs. When the mining boom passes its peak, as happened between 1960 and 1970 in the uranium industry, large numbers of people leave the area and revenues, particularly tax revenue, drop sharply. Local governments are left facing large bonded-indebtedness, and usually do not have the means to pay even the interest on this debt. This, in turn, works hardship in the form of higher local taxes, on those who remain, and those who invested face loss of interest or capital, or both. Expanded facilities go unused; unnecessary services have to be cut back, throwing people out of work. These people are forced to go elsewhere, further depleting the population and its tax revenues.

The uranium industry in Colorado, however, has been marked by only moderate growth. There were 1,008 workers in 1974; 1,040 in 1975; 1,096 in 1976, and approximately the same number in 1977. Based on past history, the socio-economic impact of the industry has probably been minimal. But, due to the recent upsurge in uranium price and demand, it is likely that the growth rate will accelerate in the near future. Fortunately, unlike the proposed development of oil shale, future uranium operations will probably be spread over ten to twelve counties rather than just three. (See Appendix E for a list of the counties with uranium operations in 1977.) Furthermore, most uranium projects will require less than 100 workers on a long-term basis. Most of the proposed oil shale, coal mine expansions, and electric-generation projects examined in the "Boom Town Financing Study" (Bolt, Luna, and Watkins, 1976) would require several hundred permanent workers by 1985. Thus, only those uranium projects that will require more than 100 permanent workers will pose potentially severe socio-economic impacts. Table 26.5 lists population estimates for those counties with uranium activity in 1977.

The counties most likely to suffer the greatest impact are those with the smallest populations: San Miguel, Saguache, Rio Blanco, Teller, and Moffat. The only "large" uranium project currently planned in one of these counties is the Pitch Project in Saguache County. The arrival of its work-force, both temporary and permanent, should give a good indication of what impact may be expected from the development of a large operation in one or more small counties.

Pitch Project manpower requirements are shown in Table 26.4. However, since many workers will bring their families with them, the total induced population could be significantly greater, as shown in Table 26.6. These figures were developed from a 1975 North Dakota State University study of coal mine and power plant workers in the Great Plains Region, and a survey of construction workers in the Craig area. Data indicated that the average family size for construction workers was 2.14 persons; for a miner, 3.47 persons; and for a mill workers, 3.97 persons (Dames and Moore, 1976). Since these family factors were applied directly to workforce numbers, they represent "worst case" figures. That is, the induced population estimates would be lower for workers not having or not bringing families. The potential construction population would also be reduced if local residents or students were hired, since they would not increase the existing community size.

The proposed Pitch mine and mill site is located on the northern edge of Saguache County, approximately 5.25 miles east of Sargents. Saguache County, population 4,114, has no towns with populations over 2,000 persons. The "town" of Sargents has fifteen year-round residents and twenty summer-only "residents." The two largest cities closest to the site are Salida (approximately thirty-five miles to the northeast in Chaffee County) and Gunnison (approximately thirty-five miles to the

TABLE 26.5

POPULATION ESTIMATES FOR COUNTIES WITH URANIUM ACTIVITY IN 1977

County	July 1, 1978 Estimates	
	Low	High
Fremont	25,800	26,600
Garfield	18,600	19,300
Grand	6,700	7,000
Jefferson	338,200	349,100
Mesa	68,400	70,700
Moffat	10,400	10,800
Montrose	21,100	21,800
Rio Blanco	5,100	5,300
Saguache	3,900	4,100
San Miguel	2,700	2,800
Teller	6,200	6,500
Weld	107,700	111,200

SOURCE: Colorado Division of Planning, 1978, p. 3.

northwest in Gunnison County). The 1976 population estimates for Salida and Gunnison were 5,158 and 5,516 persons, respectively (Lin, 1978). There is also Poncha Springs approximately 28 miles east of the site, with 339 persons in 1975. Since there is little or no housing available close to the mill site, most workers can be expected to commute from either Salida or Gunnison. This is a reasonable assumption, since some 249 miners commute from Salida to the Climax molybdenum mine, over twice the distance from Salida to the Pitch mines (Dames and Moore, 1976). Table 26.7 lists housing demand and availability in both towns, as estimated in the summer of 1977. Since 10 of the 126 preproduction workers were already residents, the net demand for 116 housing units, versus a total availability of 110 housing units, leaves a shortfall of 6 units. However, since development of the Pitch Project is currently behind schedule, no housing shortage has yet developed.

A significant housing shortage loomed in 1978, but did not occur. The workforce was expected to peak at 286 workers, before stabilizing at an estimated 206 workers, for the duration of the project. Mobile home parks in both Salida and Gunnison

TABLE 26.6

ESTIMATED INDUCED POPULATION: PITCH PROJECT

	Peak Period 1977	Winter 1977-1978	Peak Period 1978	Long Term 1978-1998
Construction Personnel				
Workers	86	25	150	---
Dependents	98	29	172	---
Total	184	54	322	---
Mine Personnel				
Workers	40	40	136	136
Dependents	99	99	336	336
Total	139	139	472	472
Mill Personnel				
Workers	---	---	---	70
Dependents	---	---	---	208
Total	---	---	---	278
Project Personnel				
Workers	126	65	286	206
Dependents	197	128	508	544
Total	323*	193	794	750

SOURCE: Dames and Moore, 1976.

*Since 10 preproduction workers were already Gunnison residents, the actual total population increase in 1977 should be no more than 288.

TABLE 26.7

ESTIMATED HOUSING DEMAND AND AVAILABILITY (Summer 1977)

	Single Family Residence	Apartment	Mobile Home	Total
Available in Gunnison	20	25	35	80
Available in Salida	15	5	10	30
Total Available	35	30	45	110
Project Pitch Total Demand	46	23	47	116*

SOURCE: Dames and Moore, 1976.

*10 of 126 preproduction workers already reside in Gunnison.

were filled and no significant new-home construction was planned as of October 1976, when the Environmental Report was released. The report recognized the problem, however, with the statement that: ". . . a serious housing shortage could be avoided by advance notice and continual communication between Pitch Project management and local government officials and developers" (Dames and Moore, 1976). The Colorado Governor's Advisory Committee (Bolt, Luna, and Watkins, 1976) recommended more involvement by energy firms in providing housing, especially for employees:

> If it does not appear that adequate housing will be available within commuting range on a timely basis, the energy company should develop alternatives for meeting housing needs. Some of the alternatives . . . to consider are: acquiring developable land; site preparation for housing construction; developing mobile home parks; and assisting builders by agreeing to purchase a number of homes at predetermined prices. The Committee does not recommend that energy firms build "company towns" except as a last resort.

Similar serious impacts on community services are not anticipated. Existing emergency services (fire and police), as well as water and sewage, should prove adequate without the addition of major capital improvements. Additionally, both Gunnison School District RE1J and Salida School District RE2J are below capacity; additional students could be absorbed—

400 in Gunnison and 274 in Salida (Dames and Moore, 1976).

The "quality of life" should remain the same. Splitting the long-term induced population of 750 between Salida and Gunnison will result in a growth rate of about 7 percent for each city, well below the 15 percent rate associated with a breakdown in local and regional institutions (U.S. Department of Housing and Urban Development, 1976). Even with growth, there will still be plenty of opportunities to "get away." Each of the three counties affected by the Pitch Project have considerable portions devoted to National Forest: Saguache (47 percent); Gunnison (60 percent); and Chaffee (68 percent). As a result, numerous outdoor activities are possible. Big game in the area include deer, elk, bear, antelope, and big-horn sheep. Detailed plans were made to insure a minimum disturbance of wildlife habitat.

Finally, the development of the Pitch Project should provide additional employment not only in the mine but also in the entire area. This will result in diversification of an economy dominated by agriculture and ranching, without disrupting the existing community structure.

An update on the Pitch Project situation reveals that, as of October 1978, the expected influx of workers and resulting housing shortage has not occurred. The Pitch Project has not obtained a mill license, and construction on the mill site has been delayed until at least spring of 1979. The workforce in 1978 has reached approximately 60 workers, with 135 more anticipated once full-scale operations begin (this estimate is down from the original estimate). This increase in population is expected to be split between Salida and Gunnison, with no significant housing shortage expected if current Homestake Mining Co. estimates and plans are correct (Thatcher, 1978). This type of anticipatory planning and assessment carried out in advance of development by the Pitch Project and similar projects can significantly reduce the negative socio-economic impacts of a large development in a sparsely populated area.

SUMMARY

Over the years, the growth of the uranium industry in Colorado has been moderate, with small mining operations predominating. However, sharp increases in the price of uranium, as well as strengthening demand, may encourage the development of large uranium operations. While uranium operations only require one-fifth to one-half the manpower of large-scale coal or oil shale operations, they can still pose serious problems, especially in smaller counties which may undergo uranium development or expansion. Actual impact will depend upon the size, timing, and location of each new uranium operation, and upon the capability of nearby communities to support housing and municipal service needs. Where communities within commuting distance of proposed operations cannot provide adequate public services and facilities, it may be appropriate for the energy firm to become actively involved. Such involvement might include builders and developers to insure adequate employee housing, as well as State and local government officials to address public service needs. Otherwise the alternative, postponement or even cancellation of a project, would be detrimental to both the community and the firm, in the long run.

27
Technological Developments Affecting Uranium Needs

PRESENT URANIUM USE

The sixty-nine nuclear power plants operating today are primarily light water reactors (LWR) (See Appendix D for a description of Colorado's Fort St. Vrain reactor), generate 12 percent of United States' electricity, and represent in-place capacity of about 50,600 MW. Approximately 185,000 MW of reactor capacity are either under construction or planned for introduction through the early 1990s, add another 142 units or 158,000 MWs of electrical generating capacity (Department of Energy, 1978c). Although nuclear electric-generating capacity is only 8 percent, 12 percent was actually produced in 1977, due in large part to the coal strike.

LWR technology is now based on using a "once-through" fuel cycle, relegating spent fuel reprocessing with recovery of unused uranium to a politically uncertain future. It is not the intent of this report to examine the controversial aspects of nuclear power use or growth, since they have been treated elsewhere (Morse, 1976). However, those issues directly related to the fuel cycle are: assurance that peaceful uses of nuclear power will not contribute to nuclear weapons proliferation; finding and providing sufficient fuel to expand nuclear electricity generating capacity in an economically and environmentally acceptable manner; and safe disposal of waste from nuclear facilities.

Each 1,000 MW of LWR capacity as presently designed and operated, requires 5,500 short tons of yellowcake (U_3O_8) over its anticipated 30-year lifetime, based on a capacity factor of 61 percent. (See Appendix G for basis of calculation.) DOE projects a domestic reserve and potential resource base of about 1.8 million tons of U_3O_8, sufficient to support the lifetime requirement for more than 600 nuclear reactors (Department of Energy, 1978c). (See also The Future, Chapter 29.)

TECHNOLOGICAL IMPACTS

The technology impacts discussed here lie within the framework of nuclear energy*: (See Appendix G for details on the Uranium Fuel Cycle.)

1. A decision to proceed with reprocessing spent reactor fuel can reduce uranium needs significantly by recycling recovered uranium and plutonium. If a decision to proceed were made today, the result would be an approximate 25 percent reduction in uranium requirements by the year 2000 (U.S. ERDA, 1976b).

2. A go-ahead on breeder reactor development (Liquid Metal Fast Breeder Reactor) could extend our projected national uranium inventory by a factor of 70 (Morse, 1976). Breeder technology is not expected to result in commercial power plants before the year 2000 even if a positive decision were made now.

3. There has been considerable focus on the front end of the nuclear fuel cycle (uranium supply) and the back end (effective waste disposal methodology) of development. Given even modest growth in nuclear reactor manufacturing, and assuming supply and waste needs are satisfied, there is increasing concern that the future fuel enrichment capacity may not prove sufficient to match reactor requirements (Bock, 1976). This concern bears directly on uranium needs.

Further, enrichment plant performance or efficiency relates directly to raw ore throughput capacity in the plant and the tails assay. The latter indicates the extent of U-235 recovery from the raw ore feed stream, presently at 0.3 percent (Figure 27.1). Reducing this to 0.2 percent would decrease yellowcake requirements by about 17 percent (Federal Energy Administration, 1976). A limiting factor in lowering tails assay is enrichment capacity, as noted earlier, and performance vs. cost tradeoffs. Enrichment capacity must be related to growth in the number of nuclear power plants.

United States enrichment facilities now use gaseous diffusion technology. Newer facilities will include gas centrifuges; these systems were proven earlier in Europe, and operate at an estimated 10 percent of the energy costs of gaseous diffusion. High speed nozzles and laser separation are also under development, with laser separation suggesting higher efficiency while using less energy.

4. Process-efficiency improvements in each step of producing uranium fuel offer potential for reducing ore needs, although major impacts are not anticipated in the foreseeable

*Those related to alternative renewable and non-renewable energy sources will not be treated in this report.

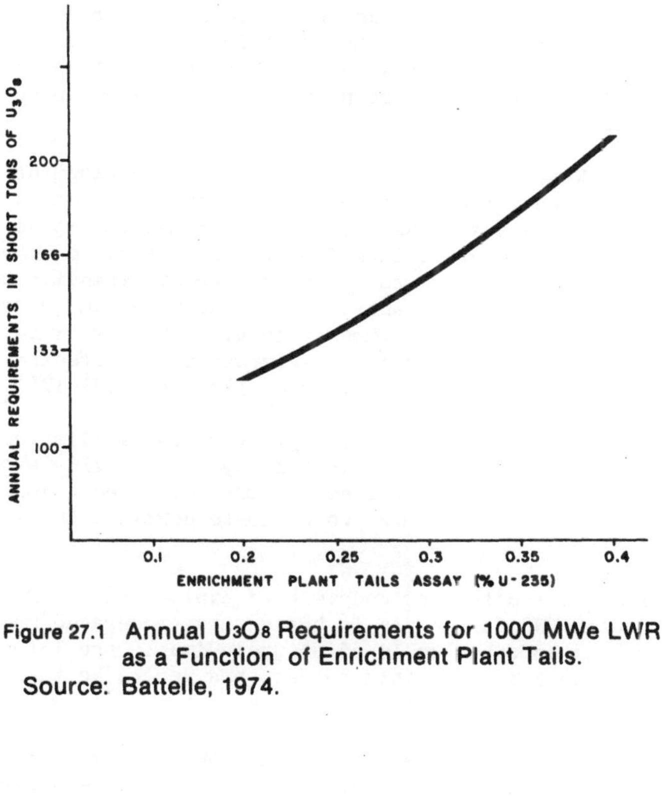

Figure 27.1 Annual U_3O_8 Requirements for 1000 MWe LWR as a Function of Enrichment Plant Tails.
Source: Battelle, 1974.

future. Beneficiation or concentration of low grade ores, prior to milling, can lead to greater recovery. Circuit modifications making use of new ion-exchange materials in milling operations can affect improved recovery of uranium from dilute solutions.

5. Development of a thorium technology base could have a larger long-term effect on uranium requirements. Thorium (Th-232), as found in nature, is a fertile material since neutron bombardment transmutes it into U-233, a fissile material. Abundance of thorium in nature is three times greater than that of uranium (U.S. Bureau of Mines, 1975b, pp. 1-2). The High Temperature Gas Cooled Reactor, such as the Public Service Company facility at Fort St. Vrain, uses thorium along with uranium in its fuel core.

The Light Water Breeder Reactor, now under development by Westinghouse in Pennsylvania, is of greater significance to uranium needs. This system uses thorium as a major component in its fuel. If successful, it offers the potential to convert present light water reactors to breeders without extensive and costly changes to existing plants. Also, the Canadian government is following these developments closely with a view to conversion of its natural uranium, heavy-water-moderated reactors to the thorium fuel cycle (Morse, 1978, pp. 157-175).

Interest stems in part from the greater availability of thorium and from the fact that this fuel cycle virtually avoids dependence upon a "plutonium economy." DOE initiated a joint effort in 1977 with the U.S.G.S. to evaluate domestic thorium resources.

6. Finally, studies are underway to explore the feasibility of reusing spent LWR fuel cores in heavy-water-moderated reactors. Such an approach would effectively double the lifetime of a uranium core; but this concept has yet to be fully evaluated in terms of cost and technology.

Since the availability of sufficient low-cost uranium is a national concern, it is possible that some of the above developments could eliminate the potential uranium shortage.

28 Uranium Production Constraints

Continued expansion of nuclear power, although of modest dimensions, appears to be politically acceptable in Washington at present. The decision is based in large part on the belief that adequate uranium supplies exist for satisfying lifetime fuel needs for over 600 LWRs (ERDA, 1977c; Ford Foundation, 1977), eliminating the need to reprocess spent reactor fuel and develop breeder reactors. Assuming the decision has a sound basis in reality, what then are the constraints? We recognize that the existence of a fuel cycle industry is essential to any expansion of nuclear power; similarly, the expansion of nuclear power will justify the existence of a fuel cycle industry.

The primary requirement for continued expansion of uranium development is the need for a responsive and responsible regulatory climate--one which encompasses federal, state, and local agencies, all necessarily involved in the decision processes. Until utilities can project the time required to obtain construction permits with reasonable accuracy, and have that time be roughly compatible with that for coal-fired plants, they will continue their reluctance to "go nuclear." On the opposite side of the issue, the reactor industry needs to develop common design standards for hardware in order to keep licensing delays at a minimum.

From the standpoint of the fuel cycle industry, uranium exploration is now operating in a "bullish" mode. For example, the October 1978 Denver Telephone Directory lists thirty-five firms under the uranium exploration category. Given encouragement by a growing number of reactor orders, fuel industry people will open more mines and build mills and enrichment plants. These plants undoubtedly will be helped by the Federal government. But mines and mills will have a common burden--acquiring financing from the private sector. All three will face the need to satisfy federal, State, and local environmental requirements.

In summary, a uranium or fuel cycle industry will continue to develop and grow, if and when serious development of nuclear power is encouraged by Washington.

29
The Future: Observations and Concerns

Public utilities, the prime users of nuclear energy, appear to be in a wait-and-see mood--waiting for federal encouragement of a more favorable regulatory climate before making a commitment to nuclear energy. It appears likely that the U.S. Department of Energy will assist in developing this favorable climate, but it is not clear how long this will take.

If the time is extended, we may see a repetition of past cyclic responses among the uranium producers. At present, producers are going forward enthusiastically, forward motion being based more on expectation than fact. One industry member has made a number of noteworthy projections for the next twelve years, some of which are listed below (Koch, 1977):

1. Exploration costs will apporach $1 billion per year by 1990.

2. Ore grade will continue its gradual decline from less than 0.2 percent to 0.12 percent.

3. Milling recovery will drop from about 95 percent as lower grade ores are processed (unless new technology intervenes. Author's note.)

4. Ore tonnage will grow 12 percent per year, increasing from 25,000 to 100,000 metric tons per day.

5. Uranium concentrate will grow 10 percent per year, from 13,000 to 40,000 metric tons per year. (Others project 80,000 metric tons per year by the year 2000 [Nuclear Fuel, 1977]).

6. By 1990 the mining rate (noted in No. 4) will require an investment ranging from $5 billion to $7.5 billion.

7. The milling rate (noted in No. 5) will require an investment of $1 billion to $1.5 billion in the same period.

8. Equipment replacement plus exploration will increase total investment to $10 billion to $15 billion in order to sustain the uranium mining and processing industry through 1990.

9. Gaseous diffusion plants will cost more than $1 billion each; centrifuge systems, although less costly, will still be expensive.

It is expected that the above will be accomplished if reactors can be constructed.

Additionally, Facer (1978) concludes:

> In 1978 we are experiencing the greatest increase in annual rates of uranium ore mining and processing during the last two decades. Mills and mines are operating at near maximum capacity, but the grade of ore mined is decreasing; therefore, it is likely that in 1979 even with three more mills and supporting mines coming into operation, the increase in production will be less than that in 1978.

References: Uranium

Ballard, T. J., and Conklin, Q. E., 1955. The uranium prospector's guide: Harper & Brothers Publishers, New York.

Battelle Pacific Northwest Laboratories, 1974. Assessment of uranium and thorium resources in the United States and the effect of policy alternatives: December, Richland, Washington.

Beckman, P., 1976. The health hazards of not going nuclear: Golem Press, Boulder, Colorado.

Beverly, R. G., 1977. A review of environmental aspects of uranium mill operations--industry's view: Union Carbide Corporation, May 24.

Bock, T., 1976. Uranium, a review of developments in the new fuel: Bosworth Sullivan and Co., Denver, Colorado.

Bolt, R. M., Luna, D., and Watkins, L. A., 1976. Boom town financing study: Vol. 1, Colorado Department of Local Affairs, Denver, Colorado.

Bowie, S. H. U., and Cameron, J., 1976. Existing and new techniques in uranium exploration, in Symposium on Exploration for Uranium Ore Deposits: International Atomic Energy Agency, Vienna.

Business Week, 1977. Issues on trial in the Westinghouse lawsuits: September 26, pp. 125-131.

Chemical & Engineering News, 1976. Adequacy of uranium supply questioned, July 19.

Chenoweth, W. L., 1977a. Uranium in western Colorado, presented at The Governor's Fourth Conference on Environmental Geology, Steamboat Springs, Colorado, September 19: Denver, Colorado.

Chenoweth, W. L., 1977b. Exploration activities, in The U.S. Department of Energy Uranium Industry Seminar, Grand Junction, Colorado, October 27, Proceedings.

Clark, D. A., 1974. State-of-the-art--Uranium mining, milling, and refining industry: Environmental Protection Agency, EPA-660/2-74-038, June, Washington, D.C.

Clayton, C. G. and others, 1976. New instruments for uranium prospecting, in Symposium on Exploration for Uranium Ore Deposits, International Atomic Energy Agency, Vienna, Proceedings.

Collins, D. B., 1977. Personal communication, November 14, 1977.

Colorado Department of Health, 1978a. Uranium mill licensing guide: Denver, Colorado.

Colorado Department of Health, 1978b. State of Colorado rules and regulations pertaining to radiation control: Denver, Colorado.

Colorado Division of Mines, 1976. Radon daughter control program: Denver, Colorado.

Colorado Division of Mines, 1978. A summary of mineral industry activity in Colorado, 1978, Part II: metal-nonmetal: Denver, Colorado.

Colorado Division of Planning, 1978. Population Estimates--August: Denver, Colorado.

Committee on Resources and Man (National Academy of Sciences), 1969. Resources and man: W. H. Freeman and Company, San Francisco, California.

Dames and Moore, 1976. Environmental report--Pitch Project, Saguache County, Colorado, prepared for the Homestake Mining Co.: Denver, Colorado.

Davis, J. F., 1977. U.S. uranium industry continues active development despite nuclear uncertainties: Engineering and Mining Journal, August, pp. 91-94.

Dodd, P. H., 1977. Uranium exploration technology, geology, mining and extractive processing of uranium, M. J. Jones, ed.: Institute of Mining and Metallurgy, London.

Dodd, P. H., and Eschliman, D. H., 1972. Borehole logging techniques for uranium exploration and evaluation, in Uranium prospecting handbook, Bowie, S. H. U. et al., eds.: Institute of Mining and Metallurgy, London.

Engineering and Mining Journal, 1977. This month in mining: September, pp. 23-33.

Environmental Protection Agency, 1978. Mining and milling administrator's guide, Contract No. 68-01-4490, Washington, D.C.

Facer, J. F., Jr., 1976. Production statistics, in The U.S. Energy Research and Development Administration Uranium Industry Seminar, Grand Junction, October: Grand Junction, Colorado.

Facer, J. F., Jr., 1977. Personal communication, November 17, 1977.

Facer, J. F., Jr., 1978. Trends in uranium production, in The U.S. Department of Energy Uranium Industry Seminar, Grand Junction, October: Grand Junction, Colorado.

Federal Energy Administration, 1976. National energy outlook: FEA-N-75/713, pp. 253-270, Washington, D.C.

Ford Foundation, 1977. Nuclear power issues and choices, in The Report of the Nuclear Energy Policy Study Group: Ballinger Publishing Co., Cambridge, Massachusetts.

Frank, J. N., 1976. Cost model for solution mining of uranium, in The U.S. Energy Research and Development Administration Uranium Industry Seminar, Grand Junction, October: Grand Junction, Colorado.

Griffith, J. W., 1967. The uranium industry - its history, technology, and prospects: Department of Energy, Mines, and Resources, Ottawa, Canada.

Grutt, E. W., Jr., 1977. Uranium exploration methods: Mining Congress Journal, April.

Hazen, W., 1978. Uranium milling, in The Uranium Resource/Technology Update Seminar, Grand Junction, July, J. G. Morse, ed.: Colorado School of Mines, Golden, Colorado.

Hebb, D., and Morse, J. G., 1976. Colorado energy resources handbook vol. I - Coal: Colorado School of Mines, Golden, Colorado.

Hetland, D. L., and Grundy, W. D., 1978. Potential resources, in The U.S. Department of Energy Uranium Industry Seminar, Grand Junction, October: Grand Junction, Colorado.

Hunkin, G. G., 1975. The environmental impact of solution mining for uranium: Mining Congress Journal, October.

Iowa Energy Policy Council, 1975. Nuclear energy--1975.

Kelly, B. T., 1974. Evaluation of a uranium deposit: Colorado School of Mines Mining Department, Golden, Colorado.

Klemenic, J., 1976. Analysis and trends in uranium supply, in The U.S. Energy Research and Development Administration Uranium Industry Seminar, Grand Junction, October: Grand Junction, Colorado.

Kober, C. L., and Proctor-Gregg, H. D., 1977. Discrimination of uranium alteration zones in selected areas by use of LANDSAT MSS imagery: U.S. Department of Energy GJBx-4(77), Grand Junction, Colorado.

Koch, L. W., 1977. Requirements in uranium mining and milling: Intermountain minerals conference: American Institute of Mining, Metallurgical, and Petroleum Engineers, July 27-30.

Lin, R., 1978. Personal communication, October 19, 1978.

McGinley, F. E., and Facer, J. F., Jr., 1976. Uranium as a by-product and by-products of uranium production: Uranium ore processing, International Atomic Energy Agency, AG/33-3, Vienna.

McGinley, F. E., 1977. Personal communication, November 11, 1977.

Malan, R. C., 1977. Personal communication, October 20, 1977.

Martin, J. B., 1977. Overview of NRC mill licensing activities, in The U.S. Department of Energy Uranium Seminar, Grand Junction, October: Grand Junction, Colorado.

Meehan, R. J., 1978. Uranium ore reserves, in The U.S. Department of Energy Uranium Industry Seminar, Grand Junction, October: Grand Junction, Colorado.

Mining Record, 1977. Cyprus announces possible uranium discovery in Colorado: September 24, Denver, Colorado.

Mining Record, 1978. Ranchers let contract for uranium leaching plant: September 20, Denver, Colorado.

Montgomery, J., 1978. Personal communication, September 9, 1978.

Morse, J. G., 1976. Nuclear power: Issues and outlook: Mineral Industries Bulletin, Colorado School of Mines, November, Golden, Colorado.

Morse, J. G., 1978. Technology developments affecting uranium needs, in The Uranium Resource/Technology Update Seminar, July, J. G. Morse, Ed.: Colorado School of Mines, Golden, Colorado.

National Institute on Occupational Safety and Health (NIOSH), 1971. Radon daughter exposure and respiratory cancer--quantitative and temporal aspects.

Nininger, R. D., 1954. Minerals for atomic energy: Van Nostrand, New York, p. 3.

Nuclear Fuel, 1977. Uranium production estimates: August 22.

Patterson, J. A., 1976. Uranium supply developments, in The Atomic Industrial Forum Fuel Cycle Conference '76, Phoenix, Arizona, March 22, Proceedings.

Root, T. E., 1976. State permits for mine development: Uranium Exploration and Development Institute, paper 11, Rocky Mountain Mineral Law Foundation.

Saum, N. M., and Lin, J. M., 1969. Exploration for uranium: Mineral Industries Bulletin, Colorado School of Mines, v. 12, no. 4, July, Golden, Colorado.

Sauvingac, N., 1978. Environmental problems of the uranium milling industry, in The Uranium Resource/Technology Update Seminar, July, J. G. Morse, Ed., Colorado School of Mines, Golden, Colorado.

Schock, D. A., 1977. Highlights of the Vail Conference on in-situ leaching of uranium: In-Situ, v. 1, no. 1.

Senftle, F. E., et al., 1976. Intrinsic germanium detector used in borehole sonde for uranium exploration: U.S.G.S. 76-452.

Shell Oil Company and Ashland Oil, Inc., 1976. Oil Shale Tract C-b socio-economic assessment, impact analysis, vol. 2, Detailed Development Plan, prepared for Area Oil Shale Office: Department of Interior, Grand Junction, Colorado.

SINB (The energy arm of the Southern Governor's Conference Energy Committee), 1976. Finance/capitalization problems of energy industries: August.

Singleton, A. H., Jr., 1968. Sources of nuclear fuel; U.S. Atomic Energy Commission, Division of Technical Information, Washington, D.C.

Slaughter, M., 1977. X-ray analysis in mineral exploration, in Nuclear methods in mineral exploration and production, J. G. Morse, Ed.: Elsevier, Amsterdam, pp. 83-87.

Smith, G. W., 1977. Status report on the development of prompt fission neutron uranium borehole logging techniques: U.S. Department of Energy GJBx-47, Grand Junction, Colorado.

Sprouse, D. P., 1976. The feasibility of entering into a uranium exploration and development program: Master's thesis T-1830, Colorado School of Mines, Golden, Colorado.

Thatcher, J., 1978. Personal communication, October 19, 1978.

U.S. Atomic Energy Commission, 1972. Leasing of AEC controlled uranium bearing lands--Colorado, Utah, New Mexico: WASH-1523, September, Washington, D.C.

U.S. Bureau of Mines, 1955. Uranium mining on the Colorado Plateau: Information Circular 7726, Washington, D.C.

U.S. Bureau of Mines, 1975a. MAS survey of uranium and vanadium deposits in Colorado (contract no. H025026): unpublished report, Washington, D.C.

U.S. Bureau of Mines, 1975b. Thorium, mineral facts and problems: Bulletin 667, Washington, D.C.

U.S. Department of Energy, 1978a. Statistical data of the uranium industry: GJO-100(78), Washington, D.C.

U.S. Department of Energy, 1978b. Survey of United States uranium marketing activity: DOE/RA-0006, May, Washington, D.C.

U.S. Department of Energy, 1978c. News release: Grand Junction Office, July, Grand Junction, Colorado.

U.S. Department of Housing and Urban Development, 1976. Rapid growth from energy projects: ideas for state and local action: HUD-CPD-140, April, Washington, D.C.

U.S. Energy Research and Development Administration, 1976a. National uranium resource evaluation, preliminary report: GJO-111(76), June, Washington, D.C.

U.S. Energy Research and Development Administration, 1976b. Benefit analysis of reprocessing and recycling light water reactor fuel: ERDA 76/121, UC-2, December, Washington, D.C.

U.S. Energy Research and Development Administration, 1977a. Statistical data of the uranium industry: GJO-100(77), January, Washington, D.C.

U.S. Energy Research and Development Administration, 1977b. Survey of United States uranium marketing activity: July, ERDA-76-40, Washington, D.C.

U.S. Energy Research and Development Administration, 1977c. A national plan for energy research, development and demonstration: ERDA 77-1, June, Washington, D.C.

U.S. Geological Survey, 1974. Nuclear energy resources: a geologic perspective: INF-74-14, Washington, D.C.

Union Carbide Corporation, 1977. News release: February 9.

University of Oklahoma, 1975. Energy alternatives - a comparative analysis: Science and Public Policy Program, Norman, Oklahoma.

Westinghouse Electric Corporation (Wyoming Mineral Corporation), 1972. Uranium solution mining: Lakewood, Colorado.

Westphal, W. H., 1978. Solution mining of uranium, in Uranium Resource/Technology Update Seminar, July, J. G. Morse, Ed.: Colorado School of Mines, Golden, Colorado.

Whicker, F. W., and Johnson, J. E., 1978. Preliminary draft for comment--radiological considerations for siting, operation, and decommissioning of uranium mines and mills: Colorado State University, Fort Collins, Colorado.

Wollenberg, H. A., 1977. Radiometric methods, in Nuclear methods in Mineral Exploration and Production, J. G. Morse, Ed.: Elsevier, Amsterdam.

Appendix D: Nuclear Energy Fundamentals

In most utility-operated power plants electricity is produced by the combustion of fuel which generates heat. Part of this heat is converted into electricity using a steam-turbine-generator set; the balance is rejected into the environment as waste heat. Heat generation drives all thermal-electric power systems. The source of heat can be the combustion of fossil fuel (coal, oil, or gas), wood, the sun's focused radiation, or nuclear fission. In general, except for the boiler design, other thermal-electric system components are similar in nature, as shown in Figure D.1.

Nuclear energy is concerned with the atom's nucleus. The nucleus is composed almost entirely of two primary particles called protons and neutrons. Each has a mass of one on the atomic mass scale. While the neutron is electrically neutral, the proton has a single positive electrical charge. Each atom is electrically neutral since its nucleus is surrounded by a cloud of electrons equal in number to the total number of protons. (An electron is a negatively charged particle of virtually negligible mass.) Atomic mass or weight is the sum of all nuclear particles. Atomic number, however, is simply the total number of protons. The distinction is necessary because the chemical identity of an atom is directly related to its atomic number, i.e., hydrogen is 1, helium 2, . . . uranium is 92.

A chemical element consists of a large number of atoms, each with the same atomic number. If we change the number of neutrons in an atom we change the atomic mass, but not the atomic <u>number</u> (nor its chemical identity). The new atom is called an isotope of the original; although its chemical properties are the same as those of the original atom, its nuclear properties may be vastly different. For example, uranium occurs in nature as three isotopes: U-234 (less than 0.01 percent); U-235 (0.7 percent); and U-238 (99.3 percent). Only U-235 is fissile; that is, it will fission when bombarded by neutrons of defined energies. U-238 and Th-232 (naturally occuring

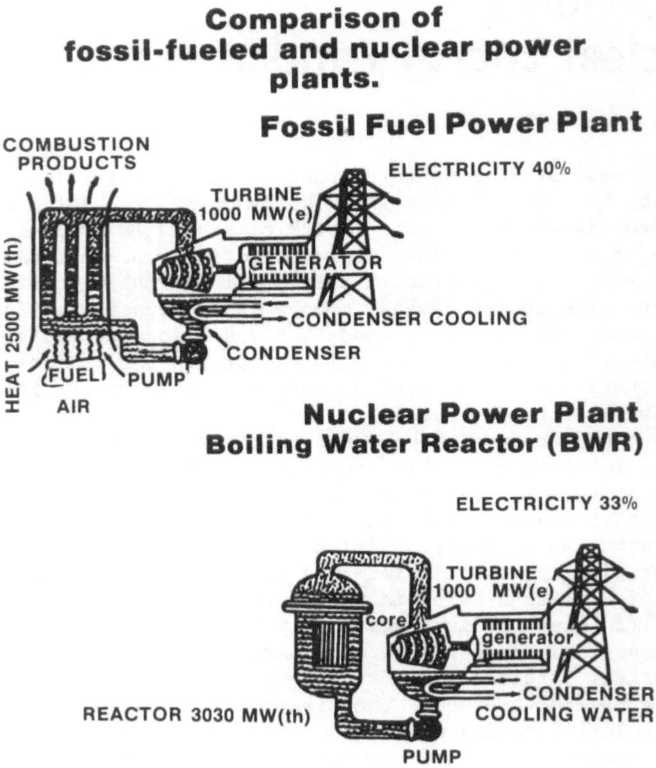

Figure D.1 Components of Fossil Fuel and Nuclear Power Plants
Source: Morse, 1976.

thorium) are fertile materials; they can be converted into fissile materials through neutron radiation, such as in a breeder-type reactor. The former converts to plutonium and the latter to uranium (U-233); both require complete chemical extraction through reprocessing for use in replacement fuel cores.

Radioactivity, or radioactive decay, is the spontaneous transformation of an atom into either a new atom or a different form of the original. In either event, the transformation is accompanied by the release of energy as alpha, beta, or gamma radiation, or some combination of these. Alpha rays are nuclei of helium atoms; they are composed of two protons and two neutrons and carry an electrical charge of +2. Beta rays are electrons which come from an atom's nucleus when a neutron decomposes to form a proton (+1) and an electron (-1). Gamma rays are pure electromagnetic radiation, like x-rays, but of higher energy. All of these are referred to as ionizing radiations because they induce electrical charges in the medium through which they pass. Their ability to penetrate matter varies considerably. For example, a small constant quantity of each type of emitter can be stopped, as follows: alphas, by a sheet of paper; betas by a Denver telephone directory; gammas, by a dozen phone books. Radiation quantity is defined in terms of the number of disintegrations, or decays, per second (dps). One curie (Ci) equals 3.7×10^{10} dps, or 37 billion dps. A more common unit, used in health and safety work is the picocurie (pCi) which is equivalent to a trillionth of a curie or 0.037 dps. (See Environmental Section for further details.)

Radiations carry off the excess energy of transformation. Each radioactive species continues its decay until it is no longer radioactive, reaching what is called its ground state. In the case of uranium, the decay process continues until non-radioactive lead is formed. Radioactive decay is a logarithmic function. In other words, if one hour is required for the radionuclide to lose half of its original activity, two hours are needed for a reduction to one-fourth, and three hours to one-eighth, etc. The term "half-life" defines the time interval necessary for half of the initial radionuclides to decay; about ten half-lives will eliminate all of the original radioactive element. The half-life of uranium is 4.5 billion years. Since this time period is approximately the same as the age of the earth, we can assume that the amount of uranium existing today is half of that at the time of the earth's formation.

Atomic power, in commercial nuclear reactors, comes from the fissioning of atoms of certain heavy elements (such as uranium-235) which split into smaller atoms, releasing large quantities of energy in the process. These smaller atoms are considerably more radioactive than the parent. Energy is produced in the form of heat, which boils water to drive the turbine, and as radiation consisting of alpha particles, beta and gamma rays, and neutrons. The fission process begins when uranium absorbs neutrons, creating an unstable species which breaks apart to achieve greater stability. Each fissioning uranium atom also emits about two additional neutrons. These "new" neutrons are absorbed in other uranium atoms, resulting in a self-sustaining chain reaction. This chain reaction may be reduced in intensity or stopped altogether by inserting control rods made of cadmium or other materials into the reactor. These rods successfully compete with uranium for the neutrons.

Three types of reactors are generally used for commercial power generation: (1) light water reactors (LWR), including both boiling water (BWR) and pressurized water reactors (PWR); (2) high-temperature gas-cooled reactors (HTGR); and (3) heavy water reactors. The United States reactor economy consists primarily of LWRs and one HTGR.

LWRs use purified tap water for two purposes: to moderate neutron energy and to carry away the heat generated within the reactor. Regarding moderation, a chain reaction can propagate within the uranium (U-235) fuel matrix as long as the neutron energies are moderated or regulated to levels which allow maximum absorption by U-235 atoms. Water performs this function; without it, the chain reaction stops. With respect to heat removal, water is converted into steam, thus carrying away the reactor heat, to drive the turbine. The two types of LWRs are described briefly below.

In BWRs, the coolant (and moderator) is vaporized to steam entirely within the reactor vessel, with reactor-generated steam driving the turbine directly, as seen in Figure D.1. In PWRs, the water circulating through the reactor core is kept in a liquid form by using high pressure. This high temperature liquid flows from the reactor through a heat exchanger, also contained in a pressure vessel, where it transfers its heat to a secondary water system. The secondary loop absorbs the heat, converting its water to steam which drives the turbine. Safe design requires operation at temperatures and pressures lower than those of fossil fired plants. Since thermal efficiency depends upon the difference in steam temperatures at the turbine inlet and outlet, PWRs have lower efficiencies than those of the BWRs and fossil-fueled plants, as noted in Table D.1.

Fuel cores for the above LWRs consist of rods packaged as bundles, containing uranium oxides enriched to about 3 percent U-235 encapsulated in non-corrosive metals.

The high temperature, gas-cooled reactor (HTGR), such as that at the Public Service Company's Fort St. Vrain plant near Platteville, Colorado, uses graphite as its moderator and helium gas as its coolant. The coolant, operating through a heat exchanger, boils water in a secondary loop to drive the turbine. This reactor functions at higher temperatures than LWRs; power plant efficiencies, therefore, are higher.

The chemically-inert helium, as the reactor coolant, will not become radioactive on exposure to the reactor's neutrons when it carries the heat from the reactor core to a secondary system--through a heat exchanger in which water converted to steam drives an efficient turbine. Further, the reactor,

TABLE D.1

POWER PLANT EFFICIENCIES (%)

Fossil Fuel	40*
PWR	31
BWR	33
HTGR	40

SOURCE: Author

*Reduces to 33-37% when scrubbers are used for SO_2 removal.

known as a "converter," uses a fuel core containing highly enriched uranium (about 90 percent in U-235) plus natural thorium. The latter is converted to uranium-233, a fissile material, through neutron absorption while in the reactor and it contributes to the fission process. Thus, it conserves the use of uranium by converting part of its fertile fuel inventory (Th-232) to fissile material (U-233).

Appendix E: Uranium Operations in Colorado

County	Name	Type Operation	Commodity	Employees
Fremont	Cotter	Mill	U, Molybdenum	112 (S)
Garfield	Rifle	Mill	V	29 (S)
	Teakee	Mine	U, V	2 (S&UG)
Grand	Big Dike Lode	Mine	U, rare earth	2
Jefferson	Foothills	Mine	U	2
	Ladwig	Mine	U	2
	Schwartzwalder	Mine	U	3
Mesa	Arrowhead #8	Mine	U, V	115 (S&UG)
	B-Chitty-U	Mine	U, V	4
	Belmont #1	Mine	U, V	3 (UG)
	Black Jack	Mine	U, V	2 (UG)
	Bonanza	Mine	U, V	3 (UG)
	Bujan	Mine	U, V	3 (UG)
	C-G-26	Mine	U, V	1 (UG)
	Economy	Mine	U, V	2 (UG)
	Hubbard	Mine	U, V	2 (UG)
	La Sal	Mine	U, V	2 (S&UG)
	Liberty Bell	Mine	U, V	3
	Lost Dutchman	Mine	U, V	3 (UG)
	Maw	Mine	U, V	3 (S)
	Mesa #5	Mine	U, V	2 (UG)
	Mineral Channel #12	Mine	U, V	2 (UG)
	New Verde	Mine	U, V	4 (UG)
	October Adit	Mine	U, V	10 (UG)
	Packrat	Mine	U, V	2 (UG)
	Peaches	Mine	U, V	6
	Rahan #30	Mine	U, V	7 (UG)
	Rajah #49	Mine	U, V	7

County	Name	Type Operation	Commodity	Employees
	Rosebud	Mine	U, V	2 (UG)
	Stafford #5	Mine	U, V	2 (UG)
	Thornton	Mine	U, V	5 (S&UG)
	Wedge	Mine	U, V	3 (UG)
	Winfield-McCormick	Mine	U, V	3 (UG)
Moffat	Maybell	Heap Leach	U	16 (S)
Montrose	Alta	Mine	U, V	4 (UG)
	Anna May	Mine	U, V	3 (UG)
	Bagger	Mine	U, V	2 (UG)
	Bitter Creek	Mine	U, V	2 (UG)
	Blackburn	Mine	U, V	3 (UG)
	Blondie	Mine	U, V	2
	Buckshot	Mine	U, V	3
	Buckhorn	Mine	U, V	7 (S)
	C-BL-23	Mine	U, V	4 (S&UG)
	C-JD-5	Mine	U, V	25 (S&UG)
	C-JD-6	Mine	U, V	16 (S&UG)
	C-JD-7	Mine	U, V	9 (UG)
	C-JD-9	Mine	U, V	7 (UC)
	C-LP-21	Mine	U, V	9 (UG)
	C-LP-22	Mine	U, V	4 (S&UG)
	Club	Mine	U, V	4 (S)
	Coloradium	Mine	U, V	3 (UG)
	Donald "L"	Mine	U, V	4 (UG)
	Duggan Adit	Mine	U, V	3 (S&UG)
	Durita Development	Leach Plant	U, V	38
	Echo #2	Mine	U, V	2 (S)
	Eula Belle	Mine	U, V	9 (UG)

County	Name	Type Operation	Commodity	Employees
	Eureka	Mine	U, V	3 (S&UG)
	Farmer Girl	Mine	U, V	2 (UG)
	Fawn Springs #9	Mine	U, V	2 (UG)
	Geo #1	Mine	U, V	5 (UG)
	King Solomon	Mine	U, V	44 (S&UG)
	Last Chance #3	Mine	U, V	4
	Lazy L	Mine	U, V	3
	Little Dick	Mine	U, V	3 (UG)
	Little Faun #1	Mine	U, V	2
	Long Park #15	Mine	U, V	3 (UG)
	Maybe	Mine	U, V	4 (UG)
	Mineral Joe #2	Mine	U, V	3 (UG)
	Mineral Joe #11	Mine	U, V	2
	Mum	Mine	U, V	3 (UG)
	Nil #2	Mine	U, V	6 (S&UG)
	Ore Buying Station	Ore Buying Station	U, V	11 (S)
	Peanuts	Mine	U, V	2 (UG)
	Peggy #2	Mine	U, V	1 (UG)
	Picket Corral	Mine	U, V	4 (UG)
	Prayer #9	Mine	U, V	2 (UG)
	Rabbit Foot #1	Mine	U, V	1 (UG)
	Radium Hill #10	Mine	U, V	3 (UG)
	Rex #38	Mine	U, V	3 (UG)
	Rim Rock #5	Mine	U, V	2 (UG)
	Rye #8	Mine	U, V	3 (UG)
	St. Patrick #7	Mine	U, V	3 (UG)
	Sandy	Mine	U, V	2 (UG)

County	Name	Type Operation	Commodity	Employees
	September Morn	Mine	U, V	2
	Showboat	Mine	U, V	3
	Silver Dollar #1	Mine	U, V	2 (UG)
	Starlight #7	Mine	U, V	4
	Sunbeam	Mine	U, V	5 (UG)
	Twilight #1	Mine	U, V	2 (UG)
	Uravan	Mine	U, V	175 (S)
	Wedge #2	Mine	U, V	2 (UG)
Rio Blanco	Midnight	Mine	U, V	3 (S)
Saguache	Pitch	Mine	U	4 (S)
San Miguel	Alice #2	Mine	U, V	2 (S)
	Ann	Mine	U, V	2 (UG)
	Brighton	Mine	U, V	3 (UG)
	Burro Tunnel	Mine	U, V	8 (S&UG)
	C-SR-16-A	Mine	U, V	4 (S&UG)
	Carnation	Mine	U, V	5 (UG)
	Civit Cat	Mine	U, V	3
	Centennial	Mine	U, V	15 (UG)
	Cougar Ventures	Mine	U, V	6 (UG)
	Deremo/Snyder	Mine	U, V	114 (S&UG)
	Dolores River	Mine	U, V	2 (UG)
	Fox	Mine	U, V	2
	Frances	Mine	U, V	2
	Helen #1	Mine	U, V	3 (UG)
	Herbert	Mine	U, V	3 (S&UG)
	Ike #1	Mine	U, V	7 (S&UG)
	Jack Knife	Mine	U, V	3 (UG)
	Letty Jones	Mine	U, V	9 (S&UG)

County	Name	Type Operation	Commodity	Employees
	Murietta	Mine	U, V	3 (UG)
	Nola Z	Mine	U, V	3
	Parrot Railroad Tunnel	Mine	U, V	2 (UG)
	Ratrick D	Mine	U, V	2
	Radium #6	Mine	U, V	2
	Riverview	Mine	U, V	2 (S)
	Sears #1	Mine	U, V	2 (S)
	Sego Lily Lou	Mine	U, V	5 (UG)
	Sheila	Mine	U, V	2 (UG)
	South Sunday	Mine	U, V	41 (S&UG)
	Suncup #2	Mine	U, V	15 (S&UG)
	Uintah	Mine	U, V	2
	Veta Mad	Mine	U, V	6 (UG)
	Windswept Group	Mine	U, V	3 (S)
Teller	High Park	Mine	U	9 (S)
Weld	Grover Test	Mine	U	22

Key: Commodity: U = Uranium
V - Vanadium

Employment: S = Surface
UG = Underground

SOURCE: Colorado Division of Mines, 1978, pp. 54-124.

Appendix F: Licensing References Pertaining to Colorado Uranium Milling

1. Title 25, Article 11, Colorado Revised Statutes, 1973, entitled "Radiation Control."

2. State of Colorado Rules and Regulations Pertaining to Radiation Control, 1978.

3. Regulatory Guide 8.11, "Applications of Bioassay for Uranium," U.S. Nuclear Regulatory Commission.

4. Regulatory Guide 8.15, "Acceptable Programs for Respiratory Protection," U.S. Nuclear Regulatory Commission.

5. Regulatory Guide 3.5, "Guide to the Contents of Applications for Uranium Milling Licenses," U.S. Nuclear Regulatory Commission.

6. Regulatory Guide 3.11, "Design, Construction, and Inspection of Embankment Retention Systems for Uranium Mills," U.S. Nuclear Regulatory Commission.

7. Regulatory Guide 4.14, "Measuring, Evaluation and Reporting Radioactivity in Releases of Radioactive Materials in Liquid and Airborne Effluents From Uranium Mills," U.S. Nuclear Regulatory Commission.

8. "Branch Position For Preoperational Radiological Environmental Monitoring Programs For Uranium Mills," U.S. Nuclear Regulatory Commission, January 9, 1978.

9. Regulatory Guide 3.8, "Preparation of Environmental Reports For Uranium Mills," U.S. Nuclear Regulatory Commission.

10. "Branch Position - Uranium Mill Tailings Management," U.S. Nuclear Regulatory Commission, May 1977.

11. The Mined Land Reclamation Act, Articles 32, Title 34, C.R.S. 1973, as amended.

12. Application For Radioactive Material License, Forms OR-RH-11 and 12, Colorado Department of Health.

13. State of Colorado Environmental Permit Directory, 1977.

14. Air Pollution Control Act, Colorado Department of Health, 1970, C.R.S. 25-7-101, et. seq.

15. "Colorado Air Quality Control Regulations and Ambient Air Quality Standards," Colorado Air Pollution Control Commission, Colorado Department of Health.

16. Subsurface Waste Disposal Regulations, Water Quality Control Commission, Colorado Department of Health, C.R.S. 24-4-103.

17. "National Pollution Discharge Elimination System (NPDES) Regulations," Water Quality Control Commission, Colorado Department of Health, C.R.S. 25-8-501, 1973 (Water Quality Control Act).

18. "Solid Waste Disposal Sites and Facilities Regulations," Division of Water Quality Engineering, Colorado Department of Health 30-2-0103, 30-20-104(3) C.R.S., 1973.

19. "Recommended Guidelines For Preparing Engineering Geologic Reports For Uranium Mill Siting, Radioactive Tailing Storage and Associated Land Use Changes," Colorado Geological Survey, Department of Natural Resources, March, 1978.

20. Colorado Historical, Prehistorical and Archaeological Resources Act of 1973.

21. Title 37, Article 87, Sections 101 through 121, C.R.S. 1973, "Reservoirs" (Department of Natural Resources, Division of Water Resources).

SOURCE: Colorado Department of Health, 1978a, pp. 25-26.

Appendix G: Uranium Fuel Cycle—Reactor Requirements

<u>Lifetime Average Annual Fuel Requirements</u>

Model 1,000 MWs light water reactor (1/3 boiling water reactor, 2/3 pressurized water reactor).

75% capacity factor (after 15th year, decrease 2% per year to 45% at year 30). 30 year plant life:

Fuel = 26.1 ST* Uranium (3.2 wt % U-235)

No plutonium or uranium recycle

Enrichment tails assay 0.25% U-235

Recoveries

93% in uranium ore processing

99.5% in conversion of concentrate to UF_6

99% in fuel fabrication

Average Ore Grade: 0.155% U_3O_8

---

*ST = Short tons (2000 lbs).

For enrichment: (Natural Uranium contains 0.711% U-235)

$$\text{Feed} = F = \overset{\text{U-238}}{\overline{0.99289}}\ F + \overset{\text{U-235}}{\overline{0.00711}}\ F$$

$$\text{Product} = P = (1.000-0.032)P + 0.032\ P$$

$$\text{Tails} = T = (1.000-0.0025)T + 0.0025T$$

$$F = P + T$$

For U-238 (1)	$0.99289\ F = 0.968\ P + 0.9975\ T$
For U-235 (2)	$0.00711\ F = 0.032\ P + 0.0025\ T$
0.968/0.032 x Eq. 2 (3)	$\underline{0.21508\ F = 0.968\ P + 0.0756\ T}$
Subtracting (3) from (1)	$0.77781\ F = 0.9219\ T$
$F = 1.1852\ T$	$= 100\%$
$P = 0.1852\ T$	$0.1852/1.1852 \times 100 = 15.63\%$ of feed
$T = F - P$	$1.0000/1.1852 \times 100 = 84.37\%$ of feed

<u>Then the annual requirement is</u>:

1. Product needed $= 26.1$ STU

2. To Fuel Fabrication $= \frac{26.1}{0.99} = 26.38 =$ Enrichment Product

3. Enrichment Feed (15.63% to Product) $= \frac{26.38}{0.1563} = 168.8$ STU in UF_6

4. Conversion Plant Feed (99.5% Recovery) $= \frac{168.8}{0.995} = 169.6$ STU in Uranium Concentrate

5. Uranium Ore Processing (93% Mill Recovery) $= \frac{169.6}{0.93} = 182.4$ STU in Uranium Ore

   U_3O_8 in Ore $= 842/714 \times 182.4 = 215$ ST U_3O_8

   For 0.155% U_3O_8 Ore

6. $\frac{215}{0.00155} = 139{,}000$ ST ore/year

These requirements are higher than those published earlier because we have assumed no reprocessing of spent fuel to recover uranium for use in reloads of fuel. Also, assumed are a lower ore grade and a lower ore processing plant recovery than others have used. The ore grade (0.155% U_3O_8) and mill recovery (93%) are the present averages for the domestic uranium industry. Reprocessing the spent fuel and recycling the uranium would reduce the requirement for uranium by about 20 percent (i.e., from 6,000 short tons U_3O_8 in concentrate to 4,900 short tons U_3O_8 for the 30-year life of a nuclear reactor).

<u>SOURCE</u>: Facer, 1977b.

Glossary

Adit - Horizontal entry into a mine.

Alpha Energy - Alpha particles are emitted from atomic nuclei with varying amounts of energy, but the energy from any radionuclide is characteristic and consistent. For example, uranium emits a characteristic 4.2 million electron volt (MeV) alpha particle and radon (Rn-222) emits a characteristic 5.5 MeV alpha particle.

Alpha Particle - A positively charged particle, composed of 2 neutrons and 2 protons, released by some atoms undergoing radioactive decay. The particle is a nucleus of a helium atom.

Ambient - A term referring to conditions in the vicinity of a reference point, usually related to the physical environment (e.g., the ambient temperature is the outdoor temperature).

Ancillary Energy - A measure of the external energy required for an energy process. It includes such things as energy for process heat, electricity for pumps, and fuel for truck, train, or barge transportation.

Anthracite - A high-rank coal with high fixed carbon, percentages of volatile matter and moisture; a late stage in the formation of coal (see "Rank"). Its energy content is about 14,000 Btu/lb.

Aquifer - Water-bearing permeable rock, sand or gravel.

Area Mining - A surface mining technique used in flat terrain.

Ash - The residue left when combustible material is thoroughly burned or otherwise oxidized.

Atomic Mass - The sum of the number of protons and neutrons in the nucleus of an atom.

Atomic Number - The number of protons in the nucleus of an atom.

Auger - A screw-type mechanism used in the transference or excavation of solid materials.

Backfilling - A reclamation technique which returns the spoils to mined cuts or pits, leaving the land in a configuration similar to the original form.

Background Radiation - The radioactivity inherent in the environment where specific radiation measurements, exclusive of the general environment, are desired. In this situation, the background radiation must be first determined and then subtracted from the total count.

Basin - A geologic or land-surface feature which is lower in the center and higher at the sides. Geologic basins may be filled with sediment and not visible from the surface.

Bench - A flat excavation.

Beneficiation - Cleaning and minimal processing to remove major impurities or otherwise improve properties.

Beta Particle - A negatively charged particle similar to an electron but emitted by the nuclei of some atoms undergoing radioactive decay. Beta particles are more penetrating but less ionizing than alpha particles.

Blanket - The area immediately surrounding the reactor core in a liquid metal, fast breeder reactor. In this case, its major function is to produce plutonium-239 from uranium-238.

Binary Cycle - Combination of two turbine cycles utilizing two different working fluids in electrical generation plants. The waste heat from the first turbine cycle provides the heat energy for the second turbine cycle.

Bitumen - A general name for various solid and semi-solid hydrocarbons; a native substance of dark color, comparatively hard and non-volatile, composed principally of hydrocarbons.

Bituminous - An intermediate-rank coal with low to high fixed carbon, intermediate to high heat content, a high percentage of volatile matter, and a low percentage of moisture (see "Rank"). About 12-15,000 Btu/lb.

Boiler - A mechanism which burns fuel to create heat energy and transfers the heat to a fluid (generally water to steam).

Breeder Reactor - A nuclear reactor that produces more fissile material than it consumes. One type is the fast breeder in which high energy (fast) neutrons produce most of the fission in current designs, and convert fertile (U-238) to fissile materials (Pu-239).

Brine - Water saturated with salt; a strong saline solution.

Btu - (British thermal unit) - The amount of heat energy necessary to raise the temperature of one pound of water by one degree Fahrenheit, specifically from 39.2 to 40.2 degrees Fahrenheit.

Box Cut - Initial excavation in a mine that penetrates a hill, resulting in walls on three sides, with spoils dumped over the slope.

Bucket-wheel Excavator - A continuous mining machine which uses scoops mounted in a circular rotating frame to remove overburden and ore deposits.

Calibration - The process of adjusting and testing the behavior characteristics of a radiation instrument, for example, so that indicated radiation intensities can be related to actual intensities.

Carnot Efficiency - The maximum efficiency with which work can be produced from heat in ideal processes. Carnot efficiency is only dependent upon the maximum and minimum temperatures available.

Carnotite - The most common secondary uranium mineral found in the Uravan mineral belt; consists of a potassium uranium vanadate.

Catalytic Conversion - A chemical reaction induced by a catalyst.

Char - A mixture of ash and carbon which remains after partial combustion or heating.

Cladding - The long tube-like container in which uranium or plutonium oxide fuel pellets are sealed.

Coal - A solid combustible organic material.

Coffinite - A black silicate uranium mineral often dispersed in and mixed with carbonaceous material.

Coke - The solid, combustible residue left after the destructive distillation of coal, crude petroleum or some other material. It is used for high-temperature metallurgical processes.

Condensation Nuclei - The small dust and aerosol particles in the atmosphere to which atomic size radon-daughters readily attach. Condensation nuclei are generally in the 0.2-0.3 micron range.

Continuous Miner - A single machine used in underground mining which accomplishes excavating, loading and initial transportation operations.

Contour Mining - A mining technique used in steeply-sloped terrain in which mining follows the seam outcrops along the side of the hill or slope.

Control Sample (of air) - Samples taken to evaluate the ventilation system and to provide information for effecting improvements. Control samples are often taken at locations where men are not working, but exposure samples are always taken at work locations.

Cracking - The process of breaking up large molecules in refinery feedstock to form smaller molecules with higher energy content.

Critical Mass - The amount of fissile material required to sustain a chain reaction.

Curie - A curie measures the radioactivity level of a substance; i.e., it is a measure of the number of unstable nuclei that are undergoing transformation in the process of radioactive decay. One curie equals the disintegration of 3.7×10^{10} or 37 billion nuclei per second.

Cyclone - A cleaning device which uses a circular flow to separate the heavier particulates from stack gases.

Decay Series - The consecutive members of a radioactive family of elements. A complete series commences with a long-lived parent, such as U-238, and ends with a stable element such as Pb-206.

Dedicated Railroad - A system in which the right-of-way, rails and rolling stock are used exclusively to transport a single resource.

Devolatilization - The removing of volatile matter from coal; mostly used as a pretreatment step to destroy the caking property of coal.

Dolomite - A mineral, $CaMg(CO_3)_2$, found as crystals in extensive beds as a compact limestone.

Dose - The amount of absorbed radiant energy. Usually given in Rems or Rads. These units are roughly 100 ergs per gram of tissue.

Dose Rate - The amount of radiant energy absorbed per unit time; usually given in Rads or Rems per hour.

Dosimeter - A device designed to indicate cumulative radiation exposure experienced by an individual. Actual absorbed energy or dose can only be inferred from the readout of the device.

Down-hole, Well Logging Instruments - Instruments which measure characteristics of formations such as electrical resistivity, radioactivity, and density. The information is used to evaluate the formations for mineral content.

Distillation - Heating a liquid mixture in order to drive off gases or vapors which are then separated according to boiling point and condensed into liquid products.

Dragline - An excavating machine used for the removal of overburden in open-pit mines. It has a boom from which is suspended a bucket which is filled by dragging.

Electrons - The orbital, negatively charged particles surrounding the nucleus of an atom.

Enrichment - The process by which the percentage of the fissionable isotope, U-235, is increased above that contained in natural uranium.

Equilibrium - The state at which the radioactive decay of consecutive elements within a radioactive series is neither increasing nor decreasing.

Equity - The net worth of a firm or corporation (total assets less total debts).

Erg - Unit of work or energy equal to about a quadrillionth (10^{-15}) of a kilowatt-hour. Work done by a force of 1 dyne acting through a distance of 1 cm.

Exothermic - Refers to a chemical reaction that gives off heat.

Exposure - The amount of radiation present in an environment, not necessarily indicative of absorbed energy but representative of potential health damage to the individual present. Working-level-hours is a good example.

Exposure Rate - The amount of environmental radiation to which an individual is subjected per unit time.

Exposure Records - Records maintained to indicate the estimated exposure of a person. Exposure records are mandatory for each employee exposed to 0.3 WL, or more, to assure that he does not acquire more than 4 working-level months per year. Normally, all major work areas occupied by each man are sampled weekly, a record of each man's occupancy time is used for calculating the necessary cumulative exposure records.

Exposure Sample - A representative radon-daughter sample, taken in a working environment, to be used in time-weighting the cumulative radon-daughter exposure of personnel.

Feedstock - Raw material supplied to a processing plant.

Fischer Assay - A standardized laboratory procedure which removes oil from oil shale, used as a basis for comparing oil shale processing alternatives and shale feedstocks.

Fission - The splitting of an atomic nucleus, resulting in the release of energy.

Fissile Material - Uranium-233, uranium-235, or plutonium-239. Fissile is a label for an atom that will fission upon absorption of neutrons.

Fixed Bed - A coal combustion (or gasification) process in which the coal is combusted on a stationary platform.

Fixed Carbon - The solid, nonvolatile, combustible portion of coal.

Fixed Cost - The cost of a business which exists regardless of the amount of production, for example, depreciation of a building or insurance.

Fixed Investment - Outlays for land, plant equipment, etc., occurring only in the initial time period of the life of an investment.

Flash Separation - Distillation to separate liquids of different volatility, accomplished by a rapid reduction in the pressure on the liquid.

Flue Gases - Gases, usually carbon dioxide, water vapor, oxides of nitrogen, and other gases, which result from combustion processes.

Fluidized Bed - A body of finely crushed particles with a gas blown through them. The gas separates the particles so that the mixture behaves like a turbulent liquid.

Fracturing - Splitting or cracking by explosion or other source of pressure to make rock permeable or loose.

Front End Loader - A tractor with a large bucket mounted on arms that can scoop up material and raise the load for dumping into a truck.

Fusion - The combining of certain light atomic nuclei to form heavier nuclei, resulting in the release of energy.

Gamma Radiation - Pure energy in contrast to particles such as beta and alpha radiation. The properties are similar to X-rays and other electromagnetic waves. Gamma radiation is highly penetrating but relatively low in ionizing potential.

Gaseous Diffusion - A process used to "enrich" nuclear fuel. The fuel, in the form of a gas, passes through a thin membrane. Light gas molecules move at a higher velocity than heavy molecules. These light molecules strike and pass through the membrane more often than the heavy molecules.

Gasification - The conversion of coal to a gaseous fuel.

Geiger Counter - A radiation detector of low efficiency used primarily for detecting total gamma radiation.

Generator - A mechanism which converts mechanical to electrical energy.

Gilsonite - Very rich tar deposits; a tar sand with a very high hydrocarbon content and low mineral content.

Half-Life - The time required for one-half of the atoms of a radioactive element to undergo decay.

Heat Exchanger - A device in which heat energy is transferred from one contained fluid to another due to a temperature difference between the two fluids.

High-Btu Gas - An equivalent of natural gas, predominantly methane; energy content is usually 950 to 1,000 Btus per cubic foot.

Highwall - The unexcavated face of exposed overburden and coal (or other resource) in a surface mine.

Hydrocarbon - Organic compounds containing only carbon and hydrogen, characteristically occurring in petroleum, natural gas, coal and bitumens.

Hydrogasification - The direct reaction of carbon with hydrogen to produce methane (CH_4).

Hydrogenation - Process of addition of hydrogen, usually to an organic compound.

HYGAS - A high Btu coal gasification process. Air heated to 800° F is blown through finely crushed coal producing an oil. Hydrogen at high pressure is injected and the mixture is heated to 1500° F to form methane.

In Situ - In the natural or original position; applied to energy resources when they are processed in the location where they were originally deposited.

Ionization - The breakdown of a molecule into its unstable charged components, consisting of either ions or free radicals. This breakdown can be caused by several methods, one of which is ionizing radiation.

Ionizing Radiation - Radiation capable of providing sufficient energy to ionize or break down molecules into charged atoms (ions).

Irradiated Fuel - Nuclear fuel that has been used in a nuclear reactor.

Isotope - One of two or more atoms with the same atomic number (i.e., the same chemical element), but with different atomic weights. Isotopes usually have very nearly the same chemical properties, but somewhat different physical properties.

Kerogen - A solid, largely insoluble organic material occurring in oil shale which yields oil when it is heated but not oxidized.

Kilocalorie - One thousand calories. A unit of energy equal to 0.004 Btu.

Leach - Removal of soluble constituents by the action of a percolating liquid.

Light Water Reactor - A nuclear reactor which uses water to (1) transfer heat from the nuclear fissioning of uranium to a steam turbine and (2) moderate or reduce the energies of the fission neutrons in order to continue the chain reaction.

Lignite - The lowest rank coal; low heat content and fixed carbon, and high percentages of volatile matter and moisture; an early stage in the formation of coal; about 6-7,500 Btu/lb.

Liquefaction of Gases - Any process by which gas is converted to the liquid state.

Liquefied Natural Gas - A clean, flammable liquid existing under very cold conditions.

Long Ton - A metric ton or 2200 pounds. A short ton weighs 2000 pounds.

Longwall Mining - Removing a mineral from an extensive exposed surface of a deposit, usually underground, by a shearing machine. Roof support is provided by movable hydraulic jacks.

Low-Btu Gas - Gas obtained by partial combustion of coal with air; energy content is usually 100 to 200 Btu per standard cubic foot.

Lurgi - A low-intermediate Btu coal gasification process in which crushed coal is heated with steam.

MMSCF - One million standard cubic feet (of gas).

Man-year - 2,080 man hours.

Megawatt - A megawatt is a million watts or a thousand kilowatts. Used as a measure of the amount of power (electricity) that can be produced by a facility at any one time, often abbreviated as MWe.

Methanation - The catalyzed reaction of CO and H_2 to form CH_4 and H_2O. Process is used to increase Btu values of low quality gas.

Methane - A colorless, odorless, flammable, gaseous hydrocarbon (CH_4), that is a product of decomposition of organic matter in marshes or mines or of the carbonization of coal. It is used as a fuel and as a raw material in chemical synthesis.

Milling - A process in the uranium fuel cycle in which ore containing only 0.2 percent uranium oxide (U_3O_8) is converted into a compound called yellowcake, containing 80 to 93 percent U_3O_8.

Mine-mouth - The vicinity or area of a mine, usually within several miles.

Moderator - A material used in thermal reactors; the purpose is to reduce the energy of neutrons to enable continuing the fission process.

Modified In-Situ - Creation of an in-situ retort by mining approximately 25 percent of the desired retort size from the bottom of the oil shale bed.

MRem - One-thousandth of a rem.

Naphtha - Any of various volatile, often flammable liquid hydrocarbon mixtures used chiefly as solvents and dilutents.

Natural Gas - A mixture of lightweight hydrocarbons in geologic deposits, with its predominant compound being methane.

Natural Background Radiation - The amount of radiation present in the environment which is not the result of man's activities.

Neutron - Electrically neutral particles found in the nucleus of all atoms heavier than hydrogen.

Nuclear Fuel Cycle - The total number of operations involved in the use of uranium in electric power production. They include: exploration, mining, milling, conversion of U_3O_8 to UF_6, enrichment, conversion of UF_6 to UO_2, fuel element fabrication, use in nuclear reactors and waste management. Spent fuel reprocessing for recovery of unused uranium and plutonium is not now being done for commercial reactor wastes.

Nucleus - The core of an atom, containing protons and neutrons and almost all of its mass.

Octane Number - A measure of a gasoline's ability to burn smoothly.

Oil Shale - Sedimentary rock (marlstone) containing insoluble organic matter (kerogen) which can be converted into oil by heating.

Once-Through Processing - The use of nuclear reactor fuel only once, avoiding reprocessing, the related problems of recovering plutonium, and nuclear weapons proliferation. This method restricts nuclear weapons proliferation and requires storage of spent fuel elements as nuclear waste.

Operating Costs - Costs that vary with the level of output such as labor costs, raw material costs, supplies, etc.

Outcrop - A site where a mineral formation is exposed to direct observation from the land surface.

Overburden - The rock, soil, etc., covering a mineral to be mined.

Permeability - The ability of a porous medium to conduct fluids through it.

Pillar - A solid mass of coal, rock, or ore left standing to support a mine roof.

Pitchblende - A complex primary ore of uranium consisting principally of uranium oxide. A massive variety of the mineral uraninite.

Picocurie - A quantitative measure of radioactivity equal to 1×10^{-12} (one-trillionth) curie. Abbreviated pCi.

Plutonium (Pu) - A heavy, man-made, fissionable, radioactive metallic element.

Possible Potential Resources - Those estimated to occur in undiscovered or partly defined deposits in formations or geologic settings productive elsewhere within the same geologic province.

Primary Containment - (Also referred to as a pressure vessel.) An enclosure surrounding the nuclear reactor core and associated equipment for the purpose of minimizing the release of radioactive material in the event of a serious malfunction in the operation of the reactor.

Primary Ventilation - The total volume of air taken underground and returned to the surface.

Probable Potential Resource - Those estimated to occur in known productive uranium districts: (1) in extensions of known deposits, or (2) in undiscovered deposits within known geologic trends or areas of mineralization.

Protons - Positively charged particles in the atomic nucleus. The number defines the chemical identity of the element.

Pyrolysis - Decomposition of materials through the application of heat with insufficient oxygen for complete oxidation.

Quad - One quadrillion Btus (10^{15} Btu's). Energy content is equivalent to 170 million barrels of crude oil.

Rad - The unit denoting absorption of 100 ergs of radiant energy per gram of absorbing material; from <u>R</u>adiation <u>A</u>bsorbed <u>D</u>ose.

Radioactivity - Spontaneous release of energy from the nucleus of an atom which may result in a change in mass.

Radiometric Prospecting - Finding minerals, using instruments such as a Geiger counter or scintillometer which measure radioactivity.

Radium - Generally refers to Ra-226, the parent of radon gas in the uranium decay series.

Radon - Normally the noble gaseous element (Rn-222) in the U-238 decay series. These include Po-218, Pb-214, Bi-214, and Po-214. They have an average combined half-life of about 30 minutes.

Rank - A classification of coal according to percentage of fixed carbon and heat content. High rank coal is presumed to have undergone more geological and chemical change than lower rank coal.

Reactor Core - The part of a nuclear power plant where fissioning occurs. It contains control rods and the fuel elements.

Rem - Radiation-equivalent man. Equals the estimated amount of energy deposited or absorbed in tissue which is biologically equivalent to 1 rad of gamma or X-radiation.

Reprocessing - The used fuel elements from a nuclear reactor are subjected to a variety of chemical and mechanical processes; the purpose is to recover the created plutonium-239 and the unused uranium-235, and to remove the fission products.

Reserves - Resources which are known in location, quantity and quality and which are economically recoverable using currently available technologies.

Resources - Total quantity of ore in the ground within specified limits of bed and overburden thickness. Includes identified and undiscovered resources.

Roentgen - A unit of radiation exposure. Technically, it is defined as that quantity of X- or gamma-radiation that produces 1 electrostatic unit of electrical charge per 0.001293 gram of air.

Room and Pillar (also room and chamber) - An underground mining technique in which areas of coal or oil shale seam are removed and columns of the deposit are left in place to support the roof.

Roll Front - A situation of non-equilibrium, i.e., uranium migration away from its highly radioactive daughters, caused by subsurface solution-dissolution processes.

Scintillation Counter - An instrument which detects activity through a crystal which emits light when contacted by radiant energy. The light pulses (scintillations) are converted to electrical pulses which are counted and correlated with the quantity of radioactivity present.

Seam - A bed of coal or other valuable mineral of any thickness.

Scrubber - Equipment used to remove pollutants, such as sulfur dioxides or particulate matter from stack gas emissions usually by means of a liquid sorbent.

Secular Equilibrium - Equilibrium where the parent of the radioactive series has a very long half-life compared to subsequent series members. An example is the U-238 series which requires about 1,000,000 years for equilibrium to develop.

Shaft - Vertical entry into a mine.

Shearing machine - An excavating machine used in longwall mining in which a rotating toothed drum cuts the seam parallel to the coal face.

Shortwall Mining - A variation of longwall mining in which a continuous miner rather than a shearer is used on a shorter working face; identical advance roof supporters are used (see "Longwall mining").

Slurry - A mixture of a liquid and solid. Explosive slurries of ammonium nitrates, TNT and water used for blasting. Slurries of oil and coal or water and coal are used in coal processing and transportation.

Specific Ionization - The number of ion pairs produced by radiant energy per linear depth of penetration. Allows a means of comparing the relative damage potential of radiations. The relative ratio of specific radiation for alpha: beta: gamma is about 100,000:100:1.

Speculative Potential Resources - These estimated to occur in undiscovered or partly defined deposits: (1) in formations or geologic settings not previously productive within a productive geologic province, or (2) within a geologic province not previously productive.

Spoils - The rock, soil, etc., of the overburden after it has been broken and removed from above the mineral seam.

Stack gas - Gases resulting from combustion.

Stream Factor - Percentage of total delivery capacity of feedstock to processor.

Subbituminous - A low rank coal with low fixed carbon and high percentages of volatile matter and moisture (see "Rank"). About 8-11,000 Btu/lb.

Subsidence - The sinking, descending or lowering of the land surface.

Sulfur Dioxide (SO_2) - One of several forms of sulfur in the air; an air pollutant generated principally from combustion of fuels that contain sulfur. A natural source of sulfur dioxide is volcanic gases.

Syncrude - A synthetic crude oil obtained by processing oil shale, tar sands or coal.

Synthane - A high Btu coal gasification process adaptable to low (lignite) through medium rank (bituminous) coals.

Synthoil - A coal liquefaction process using crushed, dried caking or non-caking coals.

Synthesis Gas - Intermediate-Btu gas; almost always used as a feedstock, but it can be used as a starting point for the manufacture of high-Btu gas, methanol or other products.

Tailings - Refuse material separated as residue in the preparation of various products (as ores).

Tar - A dark brown or black bituminous liquid obtained by destructive distillation of organic material or more commonly, a viscous oil.

Tar Sands - Hydrocarbon-bearing deposits distinguished from more conventional oil and, as reservoirs by the high viscosity of the hydrocarbon, which is not recoverable in its natural state from a well by ordinary production methods.

Thorium - A radioactive element of atomic number 90; naturally occurring thorium has one main isotope; thorium-232. The absorption of a neutron can result in the creation of uranium-233, a fissionable material.

Tritium (^3H) - A radioactive isotope of hydrogen containing 2 neutrons in its nucleus.

Turbine - A rotary engine activated by the reaction or impulse of a current or pressurized fluid (water, steam, liquid metal, etc.) and usually made with a series of curved vanes on a central rotating spindle.

Uraninite - A heavy brown, black, or dark gray mineral of uranium oxide with a pitch-like, dull, or glassy luster; a prolific source of uranium.

Uranium - A radioactive element of atomic number 92; naturally occurring uranium consists of 99.29 percent uranium-238 and 0.71 percent uranium-235.

Uranium Oxide (U_3O_8) - The most common compound of uranium found in typical ores.

Uranium Series - The 14 radioactive elements commencing with U-238 and culminating in stable Pb-206.

Vanadium - A soft, white metal, related to phosphorous; valuable as an alloy.

Vein - A fissure in a rock formation filled with mineral matter, or a bed of useful mineral.

Volatile - Readily vaporizable at a relatively low temperature.

Working Level - An atmospheric concentration of radon (Rn-222) daughters which will deliver 1.3×10^5 MeV of alpha energy per liter of air in decaying through Po-214.

Working Level Hour - An exposure equivalent to 1 working level of radon daughters for 1 hour.

Working Level Month - An exposure equivalent to 1 working level of radon daughters for 173 hours.

Yellowcake - Product of the milling process in the uranium fuel cycle that contains 80 to 95 percent uranium oxide (U_3O_8).

Index